Late Eighteenth Century Vegetation of Central and Western New York State on the Basis of Original Land Survey Records

P. L. Marks and Sana Gardescu
Ecology and Systematics
Cornell University
Ithaca, NY 14853-2701

and

Franz K. Seischab
Department of Biology
Rochester Institute of Technology
Rochester, NY 14623-0887

New York State Museum
Bulletin No. 484
ISBN 1-55557-225-1
ISSN 0278-3355

1992

The University of the State of New York
The State Education Department
The New York State Museum
Biological Survey
Albany, New York 12230

THE UNIVERSITY OF THE STATE OF NEW YORK

Regents of The University

R. CARLOS CARBALLADA, *Chancellor*, B.S.	Rochester
JORGE L. BATISTA, *Vice Chancellor*, B.A., J.D.	Bronx
WILLARD A. GENRICH, LL.B.	Buffalo
EMLYN I. GRIFFITH, A.B., J.D.	Rome
LAURA BRADLEY CHODOS, B.A., M.A.	Vischer Ferry
MARTIN C. BARELL, B.A., I.A., LL.B.	Muttontown
LOUISE P. MATTEONI, B.A., M.A., Ph.D.	Bayside
J. EDWARD MEYER, B.A., LL.B.	Chappaqua
FLOYD S. LINTON, A.B., M.A., M.P.A.	Miller Place
MIMI LEVIN LIEBER, B.A., M.A.	Manhattan
SHIRLEY C. BROWN, B.A., M.A., Ph.D.	Albany
NORMA GLUCK, B.A., M.S.W.	Manhattan
ADELAIDE L. SANFORD, B.A., M.A., P.D.	Hollis
WALTER COOPER, B.A., Ph.D.	Rochester
CARL T. HAYDEN, A.B., J.D.	Elmira
DIANE O'NEILL MCGIVERN, B.S.N., M.A., Ph.D.	Staten Island

President of The University and Commissioner of Education
THOMAS SOBOL

Executive Deputy Commissioner of Education
THOMAS E. SHELDON

Deputy Commissioner for Cultural Education
CAROLE F. HUXLEY

Assistant Commissioner for the State Museum
LOUIS D. LEVINE

Chief Scientist, Biological Survey
NORTON G. MILLER

The State Education Department does not discriminate on the basis of age, color, religion, creed, disability, marital status, veteran status, national origin, race, gender or sexual orientation in the educational programs and activities which it operates. Portions of this publication can be made available in a variety of formats, including braille, large print or audio tape, upon request. Inquiries concerning this policy of equal opportunity and affirmative action should be referred to the Department's Affirmative Action Officer, NYS Education Department, 89 Washington Avenue, Albany, NY 12234.

CONTENTS

Preface iv

Vegetation of the Central Finger Lakes Region of New York in the 1790s. P. L. Marks and Sana Gardescu.

Abstract 1
Introduction 1
Methods 1
    The study area 1
    Survey records 2
    Data on woody species 2
    Interpretation of species names 4
    Vegetation types 6
    Other boundary information 6
    Maps 6
Results 6
    Woody species 6
    Regional patterns of species distribution 15
    Minor species 18
    Forest vegetation types 19
    Openings in the forest 23
Discussion 28
    Species distributions within the region 28
    Disturbance and other open areas 30
Conclusions 32
Acknowledgments 32
Literature cited 33

Forests of the Holland Land Company in Western New York, circa 1798. Franz K. Seischab.

Abstract 36
Introduction 36
Study Area 36
Methods 37
Results 39
    Relative Frequency 39
    Relative Species Weights 40
    Community Organization 40
    Species Distributions 43
    Phytosociology and Environmental Gradients 49
    Disturbance 49
    Comparison of Forests, circa 1749-1815 51
Discussion 51
Acknowledgments 52
Literature Cited 52

PREFACE

Relatively little has been known of the forest vegetation of the northeastern United States as it existed just prior to large-scale European settlement. Several travelers' accounts describe the vegetation, but these are qualitative and not comprehensive. Notes of original rectangular land surveys provide data that allow a relatively objective picture of the vegetation to be constructed. Western New York State, a region of varied topography and geology, was one of the first areas in the United States to be surveyed in such systematic fashion.

This volume presents two studies of the late 18th century vegetation of central and western New York State. These contributions, together with the studies by Seischab (Bull. Torr. Bot. Club 117: 27-38, 1990) and Seischab and Orwig (Bull. Torr. Bot. Club 118: 117-122, 1991) provide a reasonably complete picture for most of the region.

The managing editor at the New York State Biological Survey for this New York State Museum Bulletin was Craig A. Chumbley.

Vegetation of the Central Finger Lakes Region of New York in the 1790s

P. L. Marks and Sana Gardescu

Abstract: The records made from the survey of the Military Tract in the 1790s are used to describe the vegetation that was present at that time in the central Finger Lakes region. Natural and human disturbances recorded by the surveyors, such as windfalls, burns, and cleared fields, are also reported. Woody species, forest types, and areas of open habitat are mapped at the scale of the 100 lots (each ± 1.57 km square) that made up each township, for >20 townships within the tract. Two kinds of information were used: the species of witness trees recorded at lot corners, and the surveyors' notes on trees and features of the landscape encountered along the lot boundaries.

More than 97% of the landscape was forested. Beech/maple/basswood was the predominant forest type throughout the region. Black ash swamps and other wetlands were more common in the north, on the Ontario Lowland. Oak forest was primarily in the southwest, between Seneca and Cayuga Lakes. Hemlock, cherry, and birch were common to the southeast on the Allegheny Plateau. Disturbances due to wind, fire, beavers, and people were recorded on only 1% of the lot boundaries. The windfalls occurred across the southern part of the tract, whereas burned areas were in the oak/hickory/pine region to the west. Several settlers' homes, and areas of former clearing by Native Americans, were scattered along the few roads. Other open habitats included marshes on the Lowland, and oak plains north of the Seneca River.

INTRODUCTION

Since the pioneering work of Sears (1925) and Lutz (1930), land survey records have been used repeatedly in the U.S. to convey a picture of the vegetation of a region at the time of settlement by Europeans (e.g., McIntosh 1962, Siccama 1971, Whitney 1986, Seischab 1990). Such vegetation reconstructions have been particularly valuable in areas where the vast majority of the original vegetation has been destroyed or substantially altered, making it difficult to get a sense from extant vegetation of the natural vegetation types or disturbance regime. The present paper describes the landscape in the 1790s for a part of central New York called the Military Tract, where clearing of forests for agriculture in the 1800s was extensive (e.g., Nyland *et al.* 1986, Marks and Smith 1989, Smith and Marks *in press*).

The Military Tract was created after the Revolutionary War to repay New York soldiers with grants of land and to promote settlement (Sherwood 1926). When the 28 townships within the tract were surveyed (ca. 1790-1798), the surveyors recorded the tree species growing along the boundaries and at the corners of the 100 lots in each township, and noted other features of the landscape such as swamps and windfalls.

The surveyors' records give a detailed view of the types of vegetation and disturbance found in the central Finger Lakes region 200 years ago. This information is of more than historical interest, since it improves our understanding of the origins of forest types that now occur in this region. Our study also allows a comparison of the forest types on the Allegheny Plateau and Ontario Lowland, two regions shown with differing communities on maps based on climate and soils (e.g., Bray 1930, Braun 1950). This paper complements Seischab's work, also based on survey records from the late 1700s (Seischab 1990, Seischab and Orwig 1991, and present volume), so that there is now a complete picture for much of central and western New York State.

METHODS

The study area

The Military Tract covered about 6800 km^2, from Lake Ontario south onto the Allegheny Plateau, and from Seneca Lake in the west to Oneida Lake in the east. The Tract was divided into 28 townships (Fig. 1). Relief is greater on the southeastern portion of the Tract, where some hills exceed 600 m in elevation and slopes are often >10%. The land between the Finger Lakes is level or rolling plains (Thompson 1966). To the north are the drumlins on the Ontario Lowland, where elevations are about 120 m.

In both the southern part of the Military Tract and at the northern end on Lake Ontario the soils are acidic, developed on glacial till (Cline 1970). Across the mid-region soils are more calcareous, developed on glacial till or on sediments from glacial lakes. Alluvial soils are found in stream valleys on the Plateau and on the Ontario Lowland. There are many swampy areas, both in valleys and uplands.

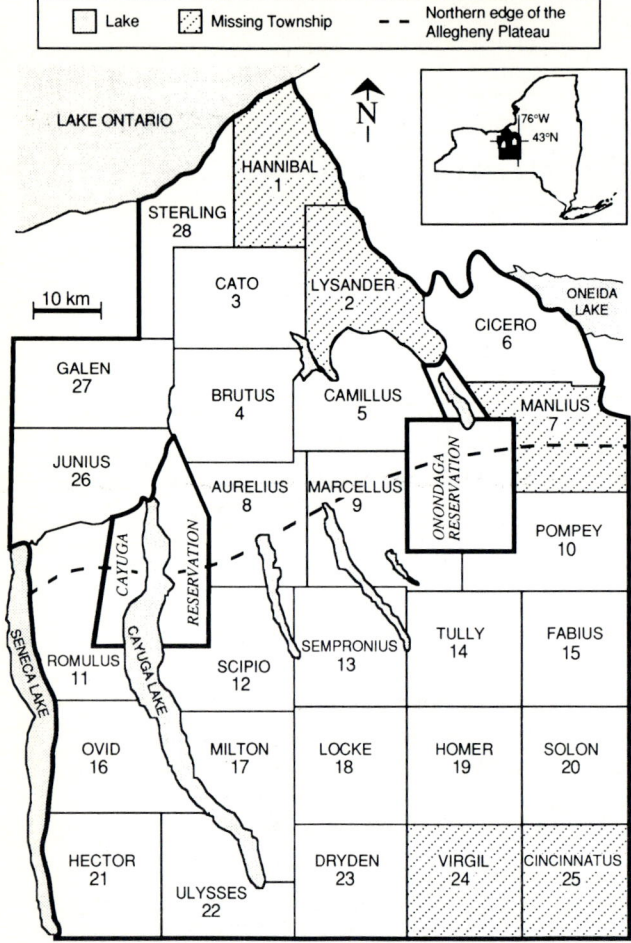

Fig. 1. Map of the 28 townships in the Military Tract. The two Indian Reservations in the region are also shown. Inset shows the location of the tract within New York State. The survey records of five townships were missing (diagonal lines). The boundary between the Ontario Lowland and the Allegheny Plateau (dashed line) is based on Fenneman (1938).

The climate is humid continental, with cold snowy winters (Thompson 1966). From the Finger Lakes north to Lake Ontario is a region of warm dry summers. To the south and southeast of the lakes, summers are cool and wet. Mean annual precipitation is about 750 to 1000 mm (Cline 1970). The average length of growing season ranges from about 135 days in the south to 180 days at Lake Ontario (Cline 1970).

Survey records

At the State Archives in Albany the senior author consulted the handwritten survey notes from the 1790s (Accession #94, vols. 24-27). All information pertinent to vegetation, topography, soils, and disturbance was read into a tape recorder, transcribed onto paper, and entered into a computer. For five townships (Table 1, Fig. 1) the survey records could not be found in the State Archives, local historical societies, or county clerks' offices. We do not know whether the records for these townships were lost or destroyed, or whether they may be preserved elsewhere. Copies of the surveyors' maps for all 28 of the townships were available at the State Archives.

Some difficulty was encountered in deciphering the handwriting of the records. The notes from one township (Solon) appeared to be the original notes written in the field at the time of survey, based on the rough handwriting, the many smudged ink spots, and the personal comments irrelevant to the survey *per se* (e.g., "Saturday, 25 June 1791: Laid still in camp for our boys to bathe."). In contrast, the perfect appearance of the notes from the remaining townships suggested that they were copied from the actual field notes. The eighteenth-century penmanship produced ambiguities (e.g., "fine" swamp vs. "pine" swamp, "butternut" vs. "bitternut"), but the number of such instances was small.

Data on woody species

Sample size

Each of the 28 townships in the Military Tract was divided into 100 lots, which were 600 acres (242.8 ha) each. Most lots were square, with the length of a side (a "bounds") slightly less than 1 mile (1.57 km; typically 78 chains, where 1 chain = 66 ft). Two kinds of information about vegetation were routinely provided by the surveyors: one witness tree at each lot corner, and brief lists of woody species encountered along the lot boundary.

In addition to the five townships for which we could not find the survey records, two (Camillus and Cicero) had witness information but no notes about boundary vegetation (Table 1, Fig. 1).

Table 1. Summary of townships, dates surveyed, and surveyors, where known. Records from five townships were not available. In Townships 5 and 6, witness trees were recorded but lot boundaries were not described. The spelling of names sometimes varied.

	Number	Township	Date	Surveyor(s)
Missing	1	Lysander	—	—
Missing	2	Hannibal	—	—
	3	Cato	—	William Ewing
	4	Brutus	—	Jacob Hart and Joseph Annin
[Witness]	5	Camillus	—	Barry Barton
[Witness]	6	Cicero	—	Boris Curtis
Missing	7	Manlius	—	—
	8	Aurelius	1790	—
	9	Marcellus	—	Jacob Hart and Joseph Annin
	10	Pompey	1791	—
	11	Romulus	—	—
	12	Scipio	—	—
	13	Sempronius	1791	Elisha Durkee
	14	Tully	1791	Jacob Hart
	15	Fabius	—	Joseph Annin
	16	Ovid	1790	Peter G. Cuddebach
	17	Milton	1790	Jacob Hart
	18	Locke	1790	Abraham Hardenbergh
	19	Homer	ca. 1790	Jacob Hart
	20	Solon	1791	Moses DeWitt
	21	Hector	1790	Sherman Nicholson
	22	Ulysses	1790	Moses DeWitt
	23	Dryden	1790-91	John Konkle
Missing	24	Virgil	1791	Moses DeWitt
Missing	25	Cincinnatus	1791	Moses DeWitt
	26	Junius	—	—
	27	Galen	1797	Joseph Annin
	28	Stirling [Sterling]	1798	Joseph Annin

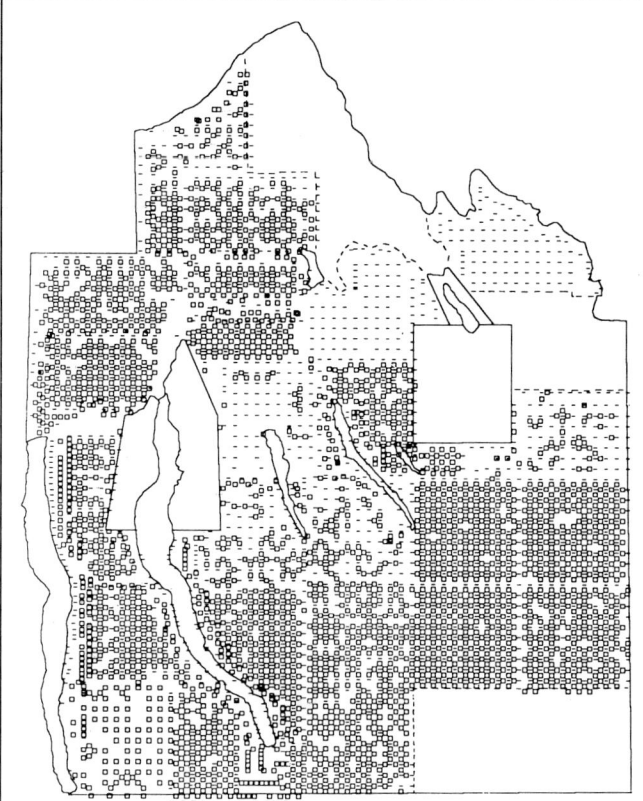

Fig. 2. Map showing which townships had more information on vegetation. Each bounds (i.e., one side of a 600-acre lot) is mapped at the center of the side of the lot. Each square represents a bounds where woody species were recorded in the surveyors' notes. Thus, where all four sides of a lot were described, they are shown as four squares. Dashes represent the witness corners where surveyors recorded the species of tree or sapling. Dashed lines mark the borders of townships with missing survey records.

Thus, out of 28 townships there were 23 for which we had witness data, covering about 5600 km^2, and 21 with boundary descriptions, or about 5100 km^2. With 100 lots per township, the potential sample size for witness trees would be about 2300. For the 21 townships with lot boundary information, there potentially could be about 4200 lot bounds, or 200 bounds per township, since the north and west sides of most lots were the south and east sides of adjacent lots. Species were not always recorded along bounds or at witness corners, and some lots were missing from the records, so the number of witness trees ranged from 12 to 118 per township, and the number of bounds from 23 to 179. However, most townships had data for >80 witness corners and >100 bounds. Total sample sizes were 1,992 witness trees and 2764 bounds in which woody species were recorded (Fig. 2).

Witness trees

Normally a single witness tree or sapling was recorded at each lot corner. Witness data were treated as presence/absence data, giving percentage occurrence of each species relative to the total number of corners at which species were recorded. A wooden stake was set at each lot corner, and some surveyors indicated the species of stake. We summarized this information on the assumption that the stakes were cut from the saplings at hand and thus may provide information about the understory.

Boundary lines

The amount of description of the boundary lines varied considerably from one surveyor to another. At one extreme, information consisted of a brief list of tree species, for example:

"Stirling [sic]... Lot 13: south bounds, heading east:
39 [chains] — brook;
72/50 [chains and links] — timber beech and maple, to a post 4 links west of a maple tree marked."

At the other extreme, detailed lists of trees were provided, sometimes understory herbs were mentioned, and major changes in species composition were noted, with the distance at each such change, as well as the beginning and end of swamps, marshes, or areas of blowdown or fire; for example:

"Fabius... Lot 7: south bounds beginning at the southwest corner, thence east:
15 — entered a swamp timbered with tamarack and black alder;
35 — out of the swamp on beech and hemlock land;
45 — a large brook running southerly and low ground covered with black alder;
50 — a brook running southerly and out of the low land into beech, maple, linden, and hemlock;
73 — set a post; marked a beech tree southwesterly 25 links. Land from the swamp good; timber as before."

Each lot bounds (i.e., each side) was treated as a line intercept sample (Seischab 1990). For each species, we used the distance along the boundaries as a measure of abundance. In the example above, tamarack occurred over a distance of 20 chains (from 15 to 35 chains). For each species, we calculated the number of lot bounds in which it was recorded and the sum of the distances of individual occurrences. These were summarized relative to the total number (2764) and total distance (4523 km) of bounds in which any woody species were recorded.

Although the majority of bounds were 1.55-1.61 km, they ranged from 0.5 to 5 km, so we also calculated the relative distance of a taxon along each bounds as a percent of that bounds' length. In the many cases where species were listed when the surveyor reached the end of the bounds, they all were given a value of 100% of bounds length. Conservatively, we attached no significance to the order of listing of species in the many cases where more than one species was recorded for a given lot boundary. Our approach differed in this regard from that used by Seischab (1990, and present volume).

Plateau vs. Lowland

To compare relative species abundance on the Allegheny Plateau and the Ontario Lowland, we mapped the location of Fenneman's (1938) physiographic boundary (see Fig. 1) onto a map of the Military Tract which shows the lots (DeWitt's State Map of New York, 1st sheet, 1792). We assigned each lot bounds and corner to one of the two regions, and for each of the common woody taxa, calculated their summed distances along lot bounds relative to the total distance within the region (3323 km for Plateau, 1192 for Lowland). Relative witness tree abundance was also summarized.

Interpretation of species names

There was some uncertainty in interpreting the surveyors' names for species of trees and shrubs. Names listed together in Table 2 indicate ones treated as synonyms for our analyses. For example, since a surveyor could enter "a black alder swamp" and then leave "the alder swamp," we treated "black alder" and "alder" as one taxon. Similarly, we grouped "white ash" with "ash," and "white pine" with "pine." We consulted a number of floras for the historical usage of common names and for the range and habitat of potential species (Torrey 1843, Paine 1865, Clute 1898, Goodrich 1912, Britton and Brown 1913, Wiegand and Eames 1926, Gleason 1952, Peattie 1966, Fernald 1970, Little 1971). Latin nomenclature follows Mitchell (1986).

Surveyors frequently used the term "maple." They also used "sugar," "sugar tree," "sugar maple," and "hard maple," which were clearly references to *Acer saccharum*. Given the distinct difference among surveyors visible at the scale of townships in the use of "maple" vs. "sugar" or "hard" maple (Figs. 3.1, 3.2), it was clear that much of the "maple" included *A. saccharum*, and therefore for most analyses these three terms were treated as a single "maple" taxon.

Surveyors also differed in number of mentions of "soft" and "white" maple, both of which were usually recorded in swamps. A reasonable interpretation is that "soft maple" (mentioned 25 times) was *A. rubrum*, which today is called soft maple as well as red maple, whereas "white maple" (used 11 times) and the one "swamp maple" referred to *Acer saccharinum*. Therefore, we treated "soft maple" as a separate taxon from "white" or "swamp" maple in the analyses. It is difficult to believe that the ecologically wide-ranging *A. rubrum* was encountered so rarely by the surveyors, as it is found today in both swamps and uplands throughout the region, so it is likely that "maple" included *A. rubrum* as well as *A. saccharum*. Perhaps the frequency of "hard" and "soft" maples recorded by the one surveyor who never used the term "maple" (Cuddebach, in the township of Ovid) reflected the relative abundance of *A. saccharum* and *A. rubrum*: only 6% were "soft." However, to keep "soft" separate from "hard," we did not include "soft maple" in our taxon group made up of "maple," "hard maple," and "sugar maple."

For most analyses the 19 occurrences of "black birch" were merged with "birch," since it was unclear how many of the 122 "birch" were *Betula lenta* (black) or *B. alleghanensis* (yellow). The term "yellow" birch was not used by any surveyor, although the species is common in the region today. "White birch," mentioned twice, could refer to *Betula populifolia* or to *B. papyrifera*. The single "paper birch" witness tree was probably the latter species, based on common names listed in Britton and Brown (1913).

The oaks were frequently not differentiated in the survey records. One surveyor recorded "black oak," "red oak," and "white oak," but only rarely "oak," but most surveyors primarily used "oak." Since even these surveyors distinguished "black" from "white" as witness trees, both species were probably included in "oak" recorded along bounds. "Black oak" was mentioned almost seven times more often than "red" oak. Was *Quercus velutina* so much more common than *Q. rubra* in 1790, unlike today? Or was "red oak" *Q. coccinea* (scarlet oak)? Because "red oak" often co-occurred with mesophytic species (e.g., maple, basswood, beech) this name would have referred to *Q. rubra*, not *Q. coccinea*. And since only three surveyors used "red," it is likely that the other surveyors included *Q. rubra* with *Q. velutina* in "black oak" and "oak." For the analysis of community types (discussed later), which can produce artificial splits among types if synonymous taxa are separated, we grouped "black," "red," and "oak." Although white oak, *Q. alba*, was probably also included in "oak," we kept "white" oak as a separate taxon because surveyors listed "black" and "white" oak as distinct species in 138 of the bounds.

We considered "scrubby" and "scrubby black" oaks to be stunted growth rather than species names, as opposed to "scrub oak," which could be *Q. ilicifolia*. Although Wiegand and Eames (1926) did not list this species in the flora of the Cayuga Lake basin, *Q. ilicifolia* was present north of Syracuse in the 1800s (Goodrich 1912).

"Whitewood" apparently referred to *Liriodendron* rather than *Tilia*, since "basswood" or "linden" were often listed in the same bounds with "whitewood." It was unlikely to refer to cottonwood, *Populus deltoides*, which can also be called whitewood (Peattie 1966), both because the surveyors did not record "whitewood" on riverbanks, and because they sometimes used the term "whitewood (popple)." Popple is a name for *Liriodendron* and the aspens (*P. grandidentata* and *P. tremuloides*), but not *P. deltoides* (Peattie 1966). The names tulip poplar, yellow poplar, and cottonwood were not used in the survey records. Use of the terms "whitewood (poplar)" and "popple alias whitewood" suggested that *Liriodendron* may sometimes have been included in "poplar" and "popple." The "poplar" on ridges with oak or chestnut would have been *P. grandidentata*. All but one of the mentions of "aspen" were by one surveyor, who never used "poplar." Since there clearly was surveyor bias in the use of these names, and since *P. grandidentata* and *P. tremuloides* were known as aspen, poplar, and popple (Peattie 1966), we treated these three synonymously (Table 2). Any mentions that included the term "whitewood," including "popple alias whitewood," we treated as *Liriodendron*. But because of this ambiguity, all of these were combined for the community analysis described below.

Habitat was a clue to species identity in several cases. Since all mentions of "spruce" were in swamps, this would have been *Picea mariana* rather than the upland species *P. rubens*. "Spruce" perhaps also included *P. glauca*, which Beauchamp (1923) found with *P. mariana* in a swamp in southern Lysander (one of the townships for which we did not have survey records). All of the "cedar" occurred in swamps, marshes, or "low" "swampy" land, and therefore would have been white cedar, *Thuja occidentalis*. The one "red cedar" would have been *Juniperus*. The "pond of lowrells" was probably bog laurel, *Kalmia polifolia*. "Huckle-berry" occurred in an area of intermixed swamps and sandy patches just north of Seneca Lake, the habitat and location reported for *Gaylussacia baccata* (Wiegand and Eames 1926).

Table 2. Names of woody species from the survey records, and the likely equivalent genera and species. Surveyors' names listed together indicate how we grouped them for analysis. Uncertain identifications are in brackets. See text for sources consulted. Latin nomenclature follows Mitchell (1986).

Surveyors' names	Likely species	
Maple	*Acer*	*saccharum, rubrum, saccharinum*
Soft maple	*Acer*	*rubrum, [saccharinum]*
White maple, Swamp maple	*Acer*	*saccharinum, [rubrum]*
Hard maple, Sugar maple, Sugar	*Acer*	*saccharum*
Alder, Black alder	*Alnus*	*incana, Ilex verticillata*
June, Juneberry, Servis	*Amelanchier*	*arborea*
Birch	*Betula*	*lenta, alleghanensis*
Black birch	*Betula*	*lenta*
White birch, Paper birch	*Betula*	*papyrifera, [populifolia]*
Hickory, Bitternut	*Carya*	*cordiformis, glabra, ovata*
Chestnut	*Castanea*	*dentata*
Dogwood	*Cornus*	*florida*
Hazel, Hazelbushes	*Corylus*	*americana, cornuta*
Thorn, Thorn bush/tree, Thorns	*Crataegus,*	*[Rosa]*
Beech	*Fagus*	*grandifolia*
Ash, White ash	*Fraxinus*	*americana, [nigra, pennsylvanica]*
Black ash	*Fraxinus*	*nigra*
Water ash	*[Fraxinus*	*pennsylvanica* or *Acer negundo]*
Huckle-berry	*Gaylussacia*	*baccata*
Butternut, White walnut	*Juglans*	*cinerea*
Walnut, Black walnut	*Juglans*	*nigra*
Red cedar	*Juniperus*	*virginiana*
Lowrell	*Kalmia*	*polifolia*
Tamarack	*Larix*	*laricina*
Whitewood	*Liriodendron*	*tulipifera*
Mulberry	*Morus*	*rubra*
Pepperidge	*Nyssa*	*sylvatica*
Ironwood, Hornbeam, Hard beam	*Ostrya*	*virginiana, Carpinus caroliniana*
Spruce	*Picea*	*mariana*
Pitch pine	*Pinus*	*rigida*
Pine, White pine	*Pinus*	*strobus, resinosa*
Buttonwood	*Platanus*	*occidentalis*
Poplar, Aspen, Popple	*Populus*	*tremuloides, grandidentata, [deltoides]*
Plum, Plumb	*Prunus*	*nigra*
Cherry, Wild cherry	*Prunus*	*serotina*
Oak	*Quercus*	*rubra, alba, velutina [montana, coccinea]*
White oak	*Quercus*	*alba*
Swamp oak, Swamp white oak	*Quercus*	*bicolor*
Scrub oak	*Quercus*	*[ilicifolia]*
Chestnut oak, Rock oak	*Quercus*	*montana*
Red oak	*Quercus*	*rubra*
Black oak	*Quercus*	*velutina, rubra*
Currants	*Ribes*	
Briers	*Rubus,*	*[Rosa]*
Willow	*Salix*	
Sassafras	*Sassafras*	*albidum*
Dogberry	*[Sorbus*	*americana, Cornus sericea,* or *Aronia arbutifolia]*
Cedar	*Thuja*	*occidentalis*
Linden, Lyn, Lime, Basswood, Bass	*Tilia*	*americana*
Swamp shuomach	*Toxicodendron*	*vernix*
Hemlock	*Tsuga*	*canadensis*
Elm, White elm	*Ulmus*	*americana, [rubra]*
Red elm	*Ulmus*	*rubra*
Cranberry	*Vaccinium*	*macrocarpon, oxycoccos*

Vegetation types

To map the distribution of forest communities in the Military Tract, and to see how the groups of species were related, the TWINSPAN program (Hill 1979) was used to produce a workable number of types from the large dataset. This program orders data from vegetation samples such that samples with similar species composition are placed close to one another. The listing of samples is repeatedly divided dichotomously, keeping samples with similar composition together. Two dendrograms showing relationships among samples are created: one for community types and one for species.

The data used in the TWINSPAN analysis were from the boundary descriptions, with each side of the lot (each bounds) treated as a single vegetation sample. For each taxon occurring in a bounds, we used its relative distance along the bounds (i.e., percent of that bounds' length) as a measure of abundance. The total sample size was 2466 bounds; bounds with incomplete distance information (e.g., in the township of Hector) were excluded from the analysis.

The TWINSPAN analysis was based on 23 woody taxa. Species mentioned in <9 bounds were excluded. Several species names used by different surveyors, which appeared to include the same taxa, were grouped in order to avoid artificial divisions of vegetation types: (a) "maple" with "hard" or "sugar" maple, (b) "oak" with "black" and "red" oak, (c) "birch" with "black" birch, and (d) "poplar" with "popple," "aspen," "whitewood," and combinations like "whitewood (poplar)."

Other boundary information

The surveyors also recorded other kinds of information about the landscape, such as swamps, marshes, windfalls, and clearings. In some bounds these were mentioned without listing any species, so the total sample size was greater than for the woody species analysis. We used 6433 km as the total distance of surveyed bounds. This was based on 4100 bounds, with a median size of 1.569 km, since each of the townships for which we had boundary descriptions had about 200 bounds (21 x 200 = 4200), but half of the records from Hector were missing. Information on swamps, disturbance, and open areas was summarized relative to these values.

Soils and topography were often described, for example "poor soil," "good land," or "ridgy." For the bounds in each of the community types produced by TWINSPAN, we compared the number of times the various descriptive terms were used. We ignored terms that clearly referred to a section of bounds that was not part of that vegetation type, for instance the comment "land between the swamps is good" for a bounds in the *Black ash swamp* type.

Maps

Many of the results were best displayed as maps of spatial distribution within the Military Tract. An arbitrary X-Y coordinate system was used, based on the surveyors' township maps (copied at a scale of 1.12 km per cm). Because witness trees were recorded only at lot corners, it was a simple matter to assign coordinates to each witness tree. However, since boundary information was recorded from near the initial corner to near the final corner and everywhere in between, mapping the boundary data was much less straightforward. The boundary data (occurrences of trees, swamps, etc.) were mapped at the midpoint of the bounds, which considerably simplified the procedure for assigning coordinates to each piece of boundary information (nearly 10,000 records) and allowed both witness and boundary information to be shown on the same map. The outlines of lakes and of the Military Tract as a whole were based on DeWitt's 1792 map (State map of New York, 1st sheet). Our maps were produced in the computer facility of the Geological Sciences Department at Cornell, using a program written by S. Gallow for the VAX computer.

To understand the relationship between the landscape and the various disturbances and open areas mentioned in the records, we marked their locations on 15-minute U.S.G.S. topographic maps. We were guided by the rivers and lakes on the surveyors' maps, by modern small roads, which sometimes follow lot edges, and for townships in Tompkins County, by a map from 1866 (Wehle 1973). At this scale, the topographic information (along with species composition and the spatial distribution of windfalls and other disturbances) allowed us to interpret the "thickets of saplings," "thick underbrush," and "scrubby" vegetation.

To show where humans were likely to have had effects on the landscape, we also mapped roads that were present in the 1790s. Some surveyors' township maps showed the positions of the roads; for other townships we used the boundary descriptions, and mapped the locations where surveyors recorded roads that intersected lot bounds. The usual term used was "road," but there were also a few mentions of "Indian foot path," "horse road," and "cart road." In townships where surveyors did not map any roads, and where roads were recorded in the descriptions of certain bounds but not on the adjacent lots, we used two maps by Simeon DeWitt (Map 103C in Cook 1887, and the 1st sheet of the 1792 state map) to fill in the gaps. DeWitt's Map 103C was based on a surveyor's report made during General Sullivan's military campaign through this region in 1779. His 1792 state map includes a road that may post-date the survey, since it appears to run along township and lot boundaries, and was not mentioned in the survey notes.

RESULTS

Woody species

More than 50 species of trees and shrubs were mentioned by the surveyors in the course of describing 2764 lot bounds (Table 3). Surveyors usually listed only two to four species in each bounds (75% of the time); more than 6 species were listed in only 2% of the bounds, and the maximum was 11 co-occurring species. Thirty-six taxa were recorded as witness trees in a total of 1,992 lot corners (Table 3). Several of the entries in Table 3

Table 3. Abundances of woody taxa mentioned in the survey notes, along lot bounds and at witness corners. Line divides the common taxa (recorded in >15 bounds) from minor species (in <9). Two measures of bounds abundance are shown: the summed distances along which the taxon was recorded, relative to the total distance of all bounds (4523 km), and the number of bounds, relative to the total number (2764). These totals exclude bounds where the surveyors did not record any woody species. Percents do not sum to 100 because species co-occurred. For witness corners, abundances are relative to the total number of corners with species recorded; percents sum to 100.

Species	BOUNDS		WITNESS CORNERS	
	Distance (% of 4523 km)	Number (% of 2764)	Trees and saplings (% of 1992)	Stakes (% of 483)
Beech	60.7%	72.0%	46.5%	36.6%
Maple	41.9	51.9	11.4	5.4
Linden/Basswood	41.0	47.3	3.9	6.6
Hard maple/Sugar maple	15.1	15.7	8.3	16.6
Hemlock	12.9	19.2	4.9	1.2
Ash/White ash	11.3	14.0	2.5	3.5
Elm/White elm	10.8	13.4	2.0	3.5
Oak	8.8	10.7	0.05	0.6
Hickory/Bitternut	6.8	7.9	1.7	5.6
White oak	6.5	7.9	4.3	0.4
Pine/White pine	6.3	9.6	1.4	1.0
Black oak	5.1	5.6	1.9	1.4
Chestnut	3.9	5.5	0.6	1.2
Cherry	3.6	4.3	0.2	0.2
Black ash	2.9	12.1	3.2	0.4
Birch	1.9	2.7	2.3	0.4
Butternut/White walnut	1.7	2.0	0.4	0.6
Alder/Black alder	0.8	3.0	0.05	0.6
Walnut/Black walnut	0.8	1.0	.	.
Ironwood/Hornbeam	0.5	0.7	1.6	10.4
Red oak	0.5	0.9	0.2	0.2
Poplar/Aspen/Popple	0.5	0.9	0.6	0.4
Cedar	0.5	1.3	.	.
Whitewood	0.4	0.6	0.1	0.2
Tamarack	0.4	1.6	0.3	0.2
Pitch pine	0.3	0.6	0.3	.
Black birch	0.2	0.6	0.2	.
Soft maple	0.2	0.7	0.3	0.4
Swamp oak/Swamp white oak	0.1	0.2	.	.
White maple/Swamp maple	0.1	0.3	0.3	.
Thorn/Thorns	0.1	0.3	.	0.2
Scrub oak	0.07	0.3	.	.
Dogwood	0.07	0.1	.	0.6
Willow	0.06	0.2	0.05	.
Briers	0.05	0.2	.	.
Spruce	0.05	0.2	.	.
Hazel/Hazel bushes	0.04	0.07	.	.
Chestnut oak/Rock oak	0.04	0.1	0.1	.
Plum	0.04	0.07	.	.
White birch/Paper birch	0.03	0.07	0.05	.
Currants	0.03	0.1	.	.
Buttonwood [i.e., sycamore]	0.03	0.1	0.3	.
Huckle-berry	0.03	0.07	.	.
Sassafras	0.01	0.04	.	0.2
Cranberry	0.01	0.1	.	.
June/Juneberry/Servis	0.01	0.04	0.2	0.6
Water ash	0.007	0.04	.	.
Pepperidge	0.001	0.04	0.05	.
Lowrells [i.e., laurel]	0.0004	0.04	.	.
Mulberry	0.0004	0.04	.	.
Red elm	.	.	0.1	0.2
Red cedar	.	.	0.05	.
Swamp shuomach	.	.	0.05	.
Dogberry	.	.	.	0.2

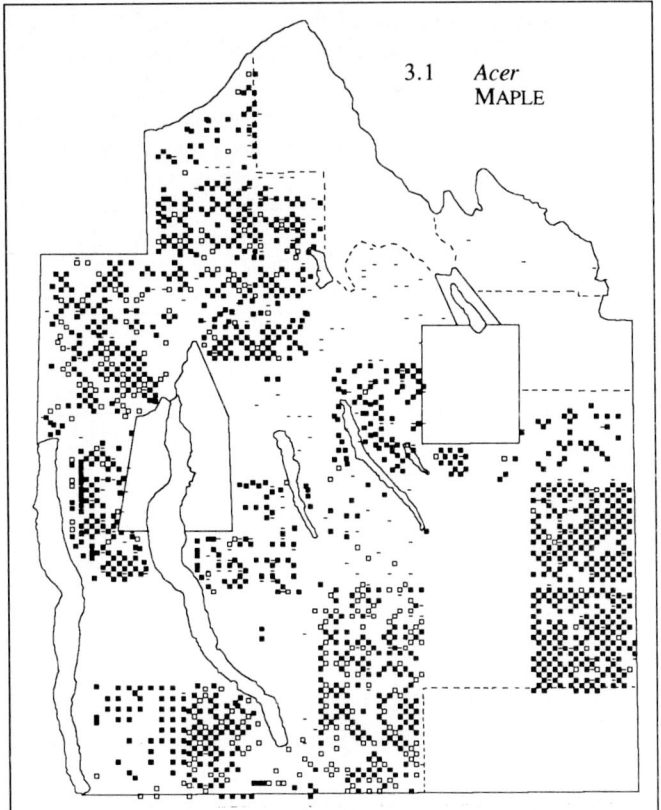
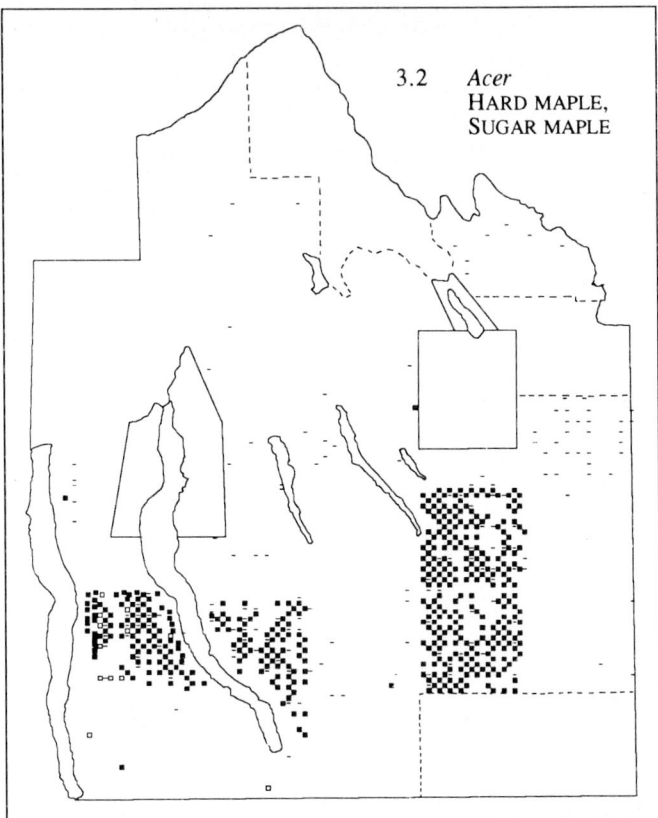

Figs. 3.1-3.27. Maps of the occurrences of the common woody taxa in the Military Tract, as recorded by the surveyors. The taxa are ordered alphabetically by genus (as in Table 2). Dashes represent witness trees or saplings at the lot corners. Boundary information is mapped at two levels of abundance: open squares represent bounds where taxa were recorded along <50% of the length; filled squares, ≥50% of the bounds length. Note that apparent gaps in species distributions should be compared with Figure 2 to see if these are areas where no species were recorded. For comparison with soft maple, "white" and "swamp" maple are also shown (same symbol for bounds or witness).

represent more than one species: "ironwood" presumably included both *Carpinus* and *Ostrya*, and "hickory" would have included bitternut, shagbark and pignut; only one "bitternut" was mentioned by name. Half of the taxa were mentioned <15 times (Table 3); these minor species are discussed later.

Geographic distributions of the occurrences of tree taxa mentioned in the survey notes are given in Fig. 3. Beech was the most common woody species encountered by the surveyors (Table 3, Fig. 3.8). Beech occurred in 72% of the lot bounds, and along 61% of the total distance (with other species or alone). Beech was also the most abundant species as witness tree or as wood for witness corner stakes.

The rank order of species based on the number of bounds was reasonably congruent with the order based on total boundary distance (Table 3). The exceptions, such as black ash, recorded in 12% of the bounds but only 3% of the distance, were taxa that often occurred on only a small distance along individual bounds. Taxa with the lowest relative bounds lengths (Fig. 4) were all swamp species, including black ash, alder, and tamarack (Table 4 lists species recorded in swamps). Wetland taxa were seldom recorded along entire lot bounds; such bounds were either described as swamps without mentioning species, or were not traversed (e.g., "through bad swampy land to a marsh, which I left being impracticable to pass"). The median distance in swamp, for bounds in which it was recorded, was quite small: 0.24 km (Table 4). Since most bounds were 1.6 km, this created the differences in relative bounds number and relative total distance (Table 3) for the species found mainly in swamps. Similarly, species of dry ridges such as chestnut and pitch pine also had low relative bounds lengths (Fig. 4). In contrast, virtually all of the common upland species, in bounds where they were listed, were usually recorded for the entire length of the bounds (medians = 100%; Fig. 4).

For the lot boundary data, beech, maple (presumably mostly sugar maple), and linden (i.e., basswood) were overwhelmingly the most abundant species, whether based on occurrence or distance (Table 3, Figs. 3.8, 3.1, 3.25). The same was not true for the witness data. In the boundary data, maple and hard maple together were nearly as abundant as beech, whereas in the witness data beech was more than twice as abundant as maple plus hard maple; 47% vs. 20% (Table 3). No other species exceeded 5% of the witness trees: hemlock was third in abundance at 4.9% and white oak was fourth at 4.3%. Only a single "oak" witness tree was recorded, although it was common on bounds; as mentioned above, at lot corners the surveyors specified oaks as white or black. The total frequencies for all species of oak were 6.5% of the witness trees and 19.4% of the number of bounds, similar to hemlock (4.9% of witness, 19.2% of bounds). Only 3.9% of the witness trees were linden, perhaps reflecting surveyor bias, a more clumped distribution than for beech and maple, or a lower

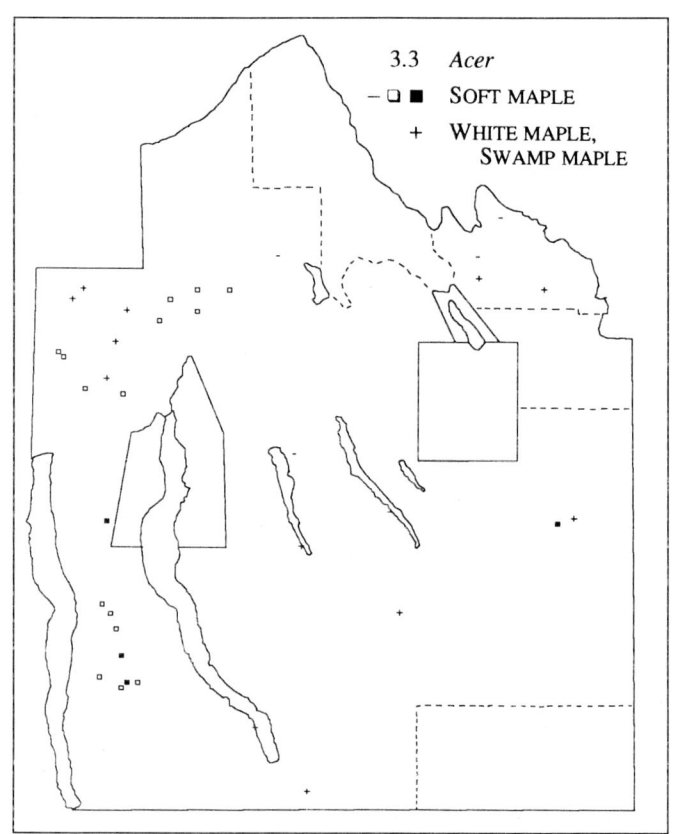
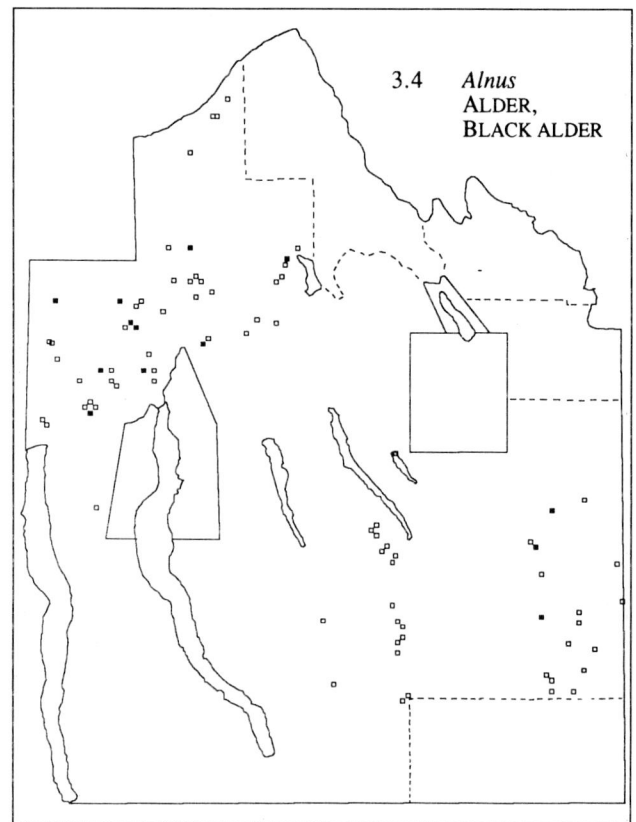
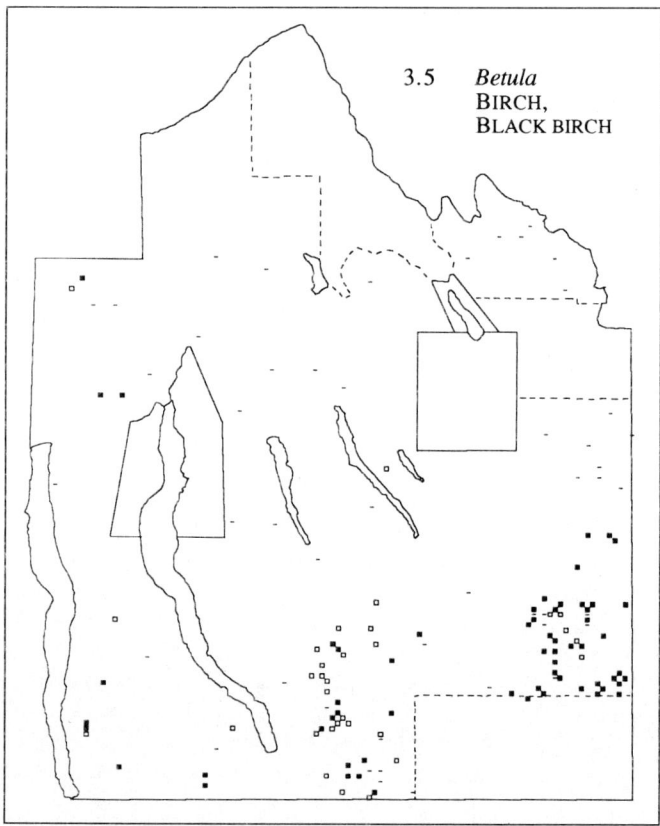
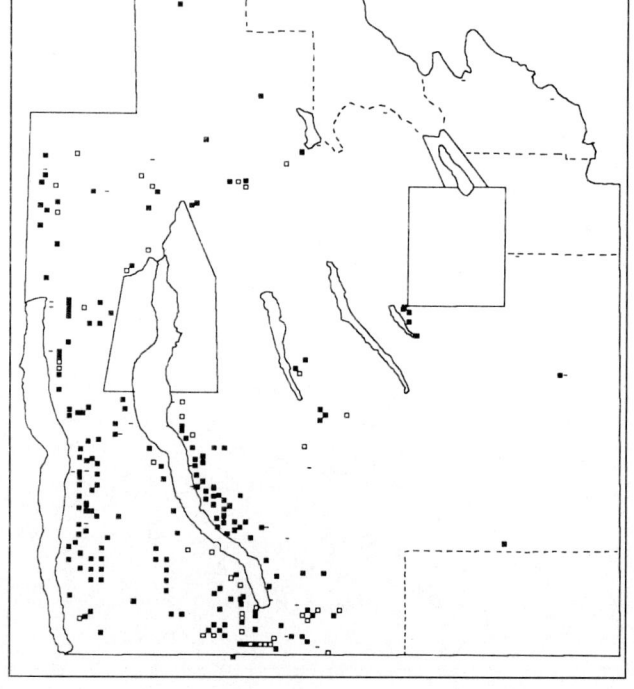

Fig. 3 (continued)

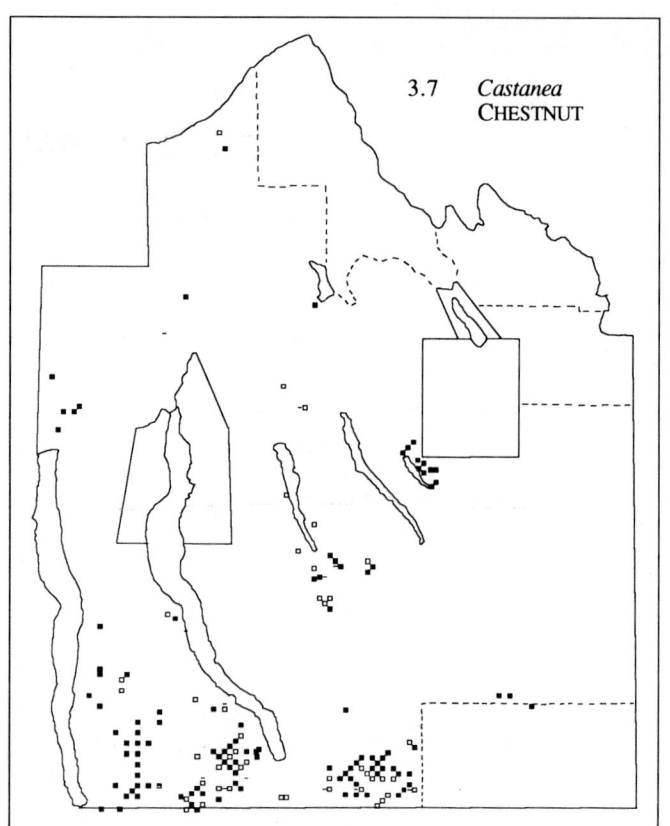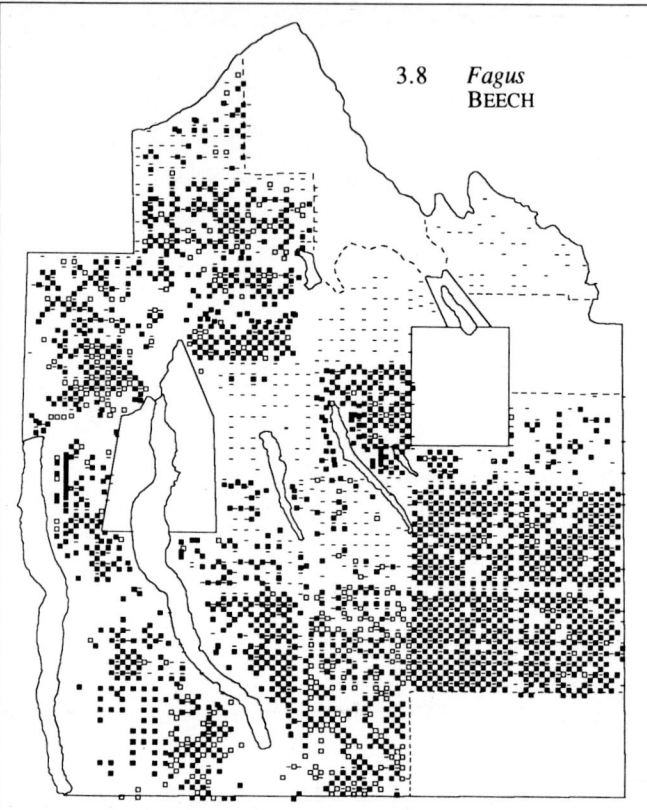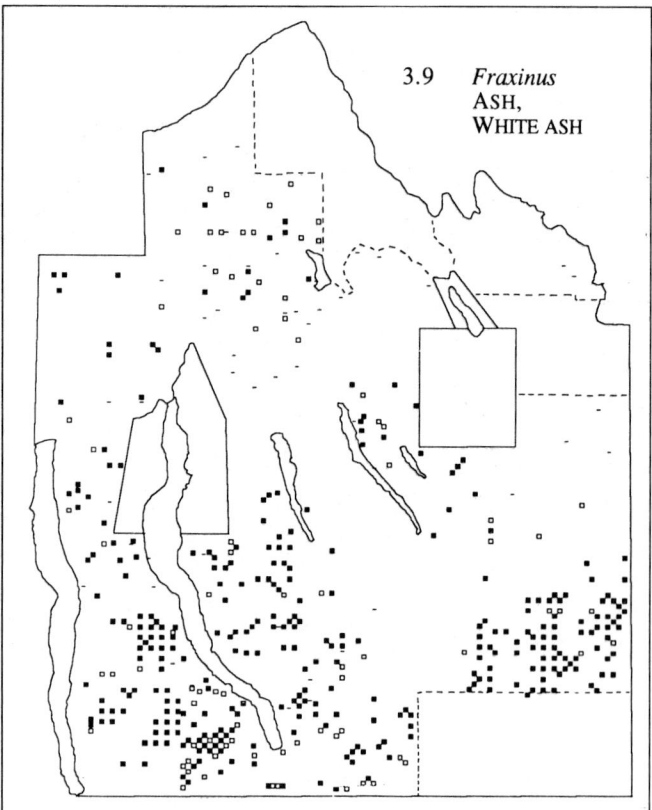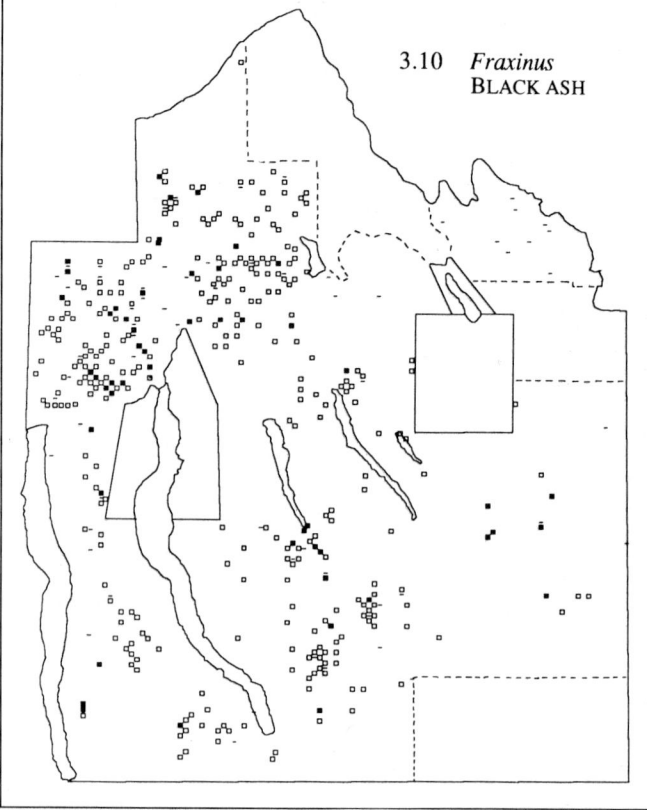

Fig. 3 (continued)

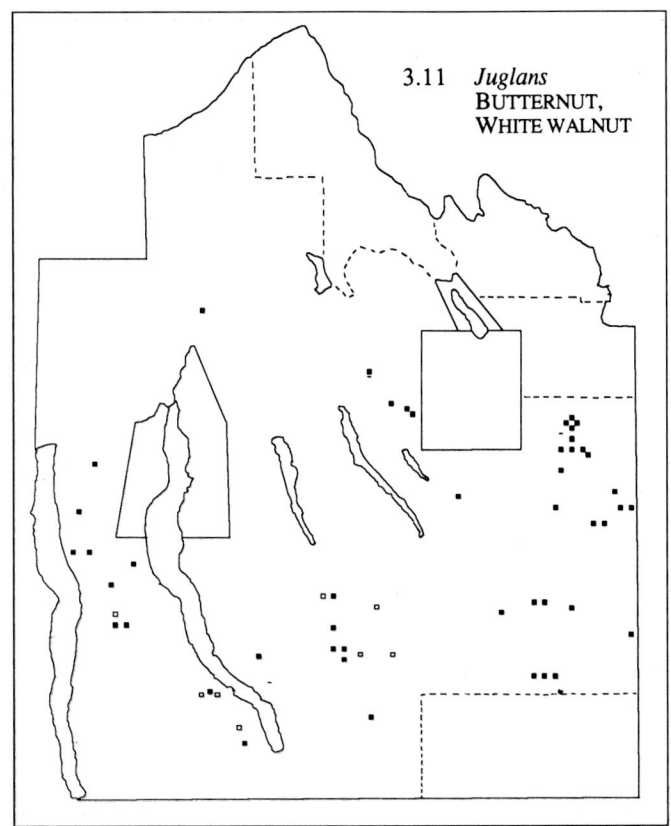
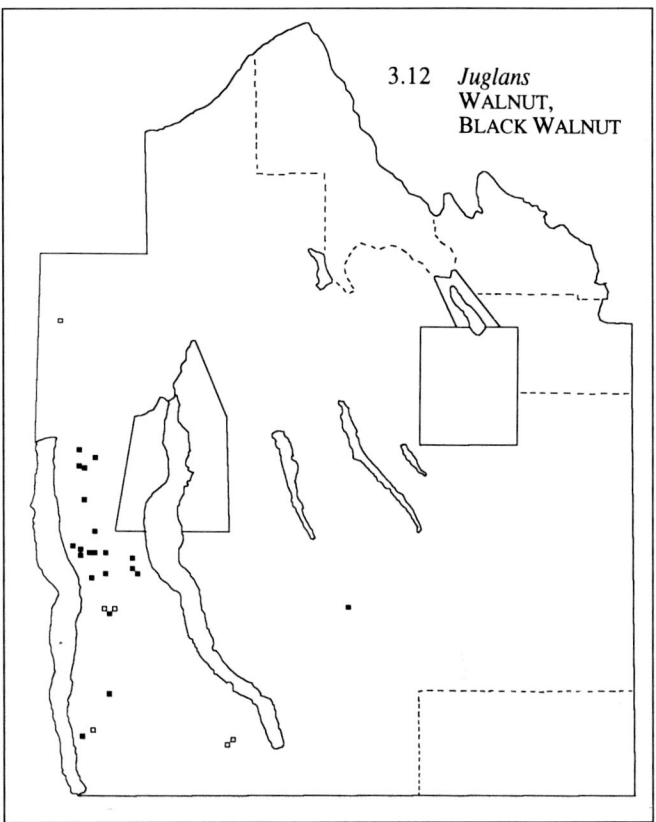
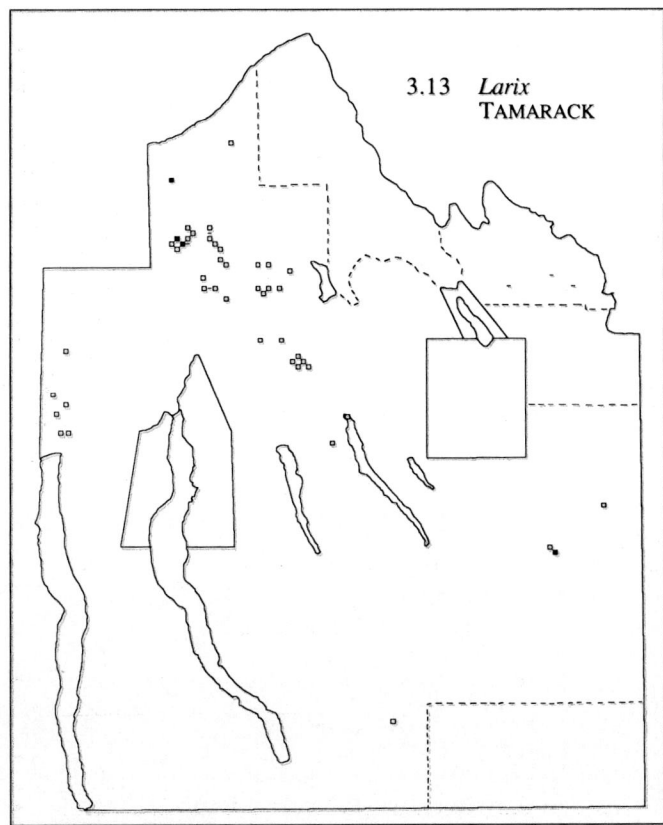
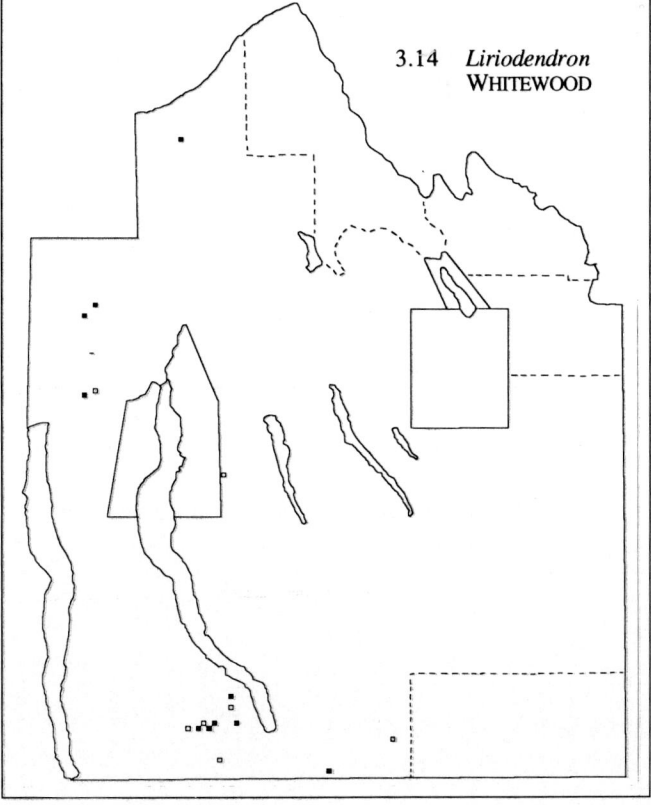

Fig. 3 (continued)

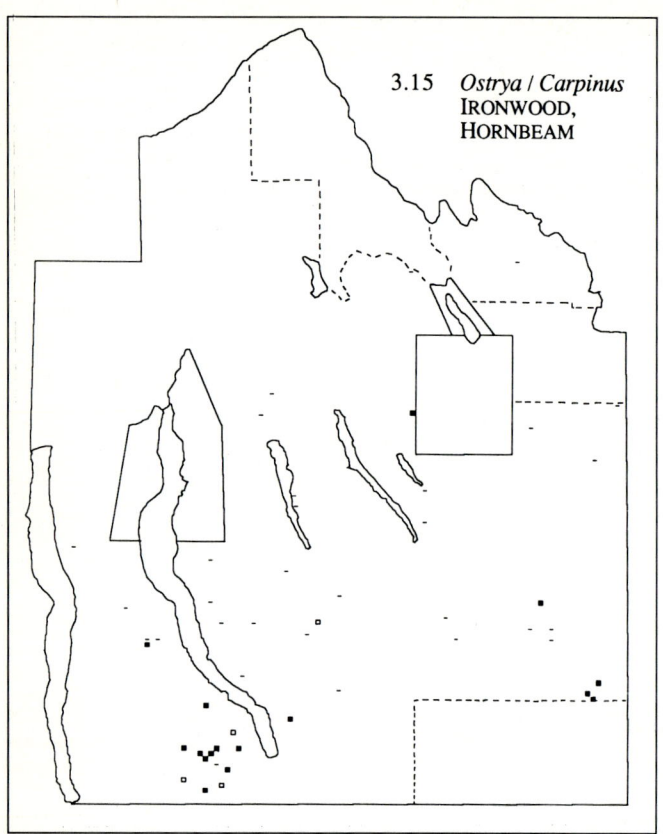
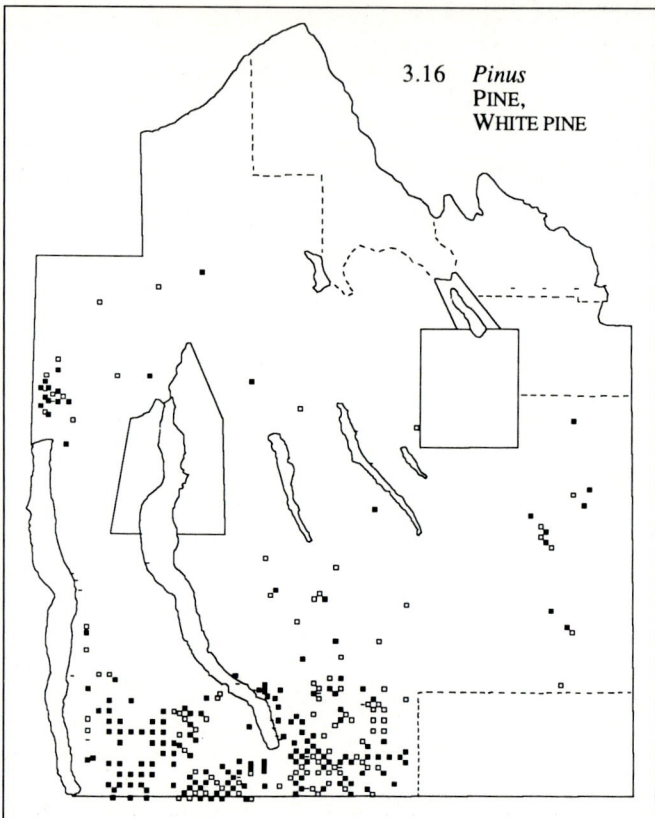
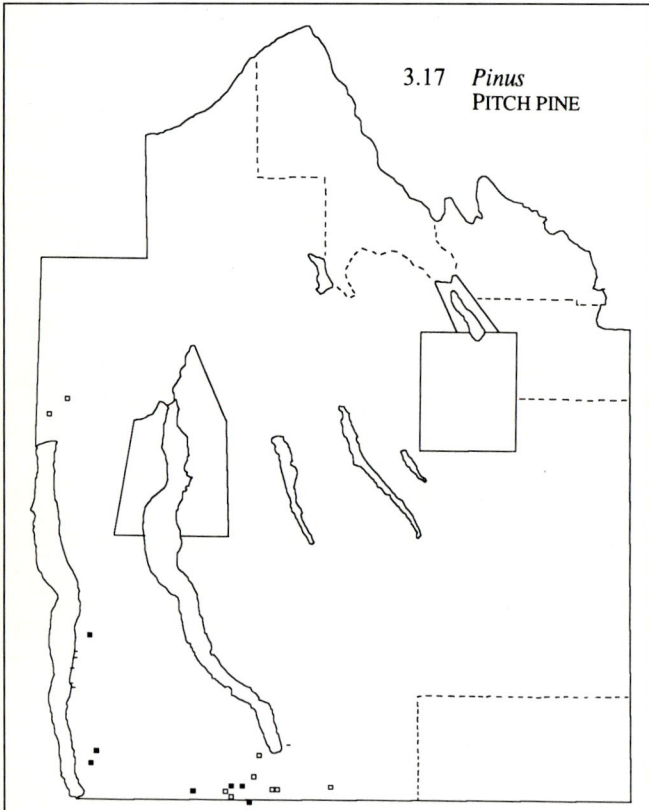
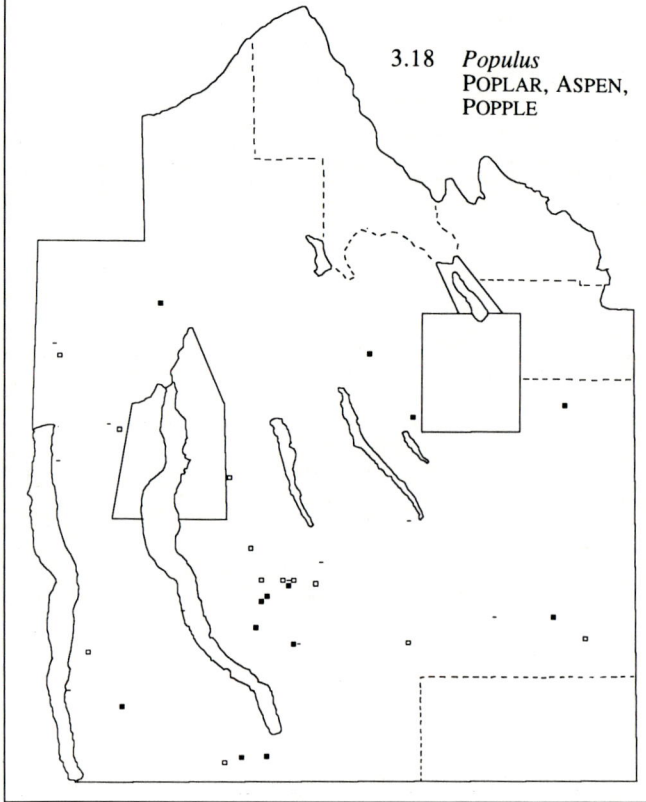

Fig. 3 (continued)

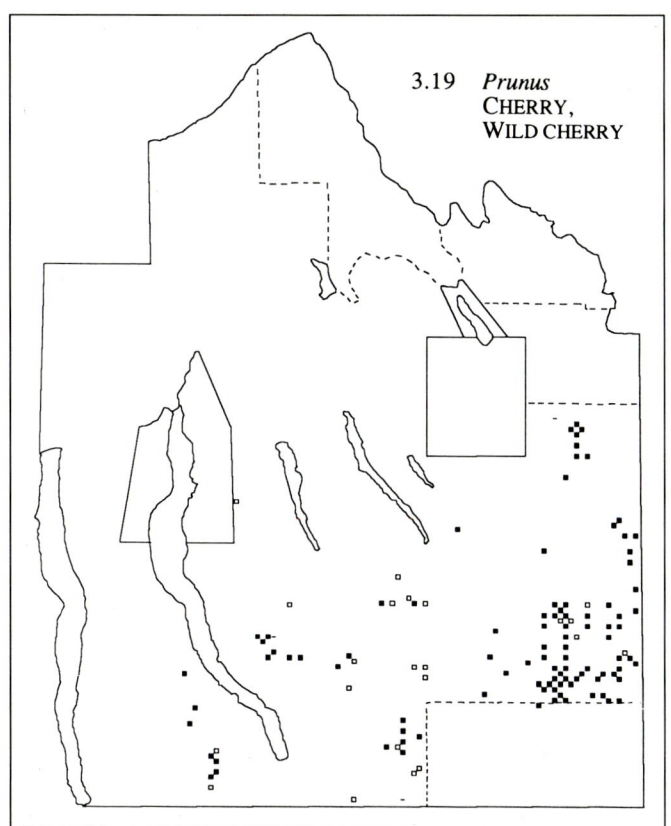
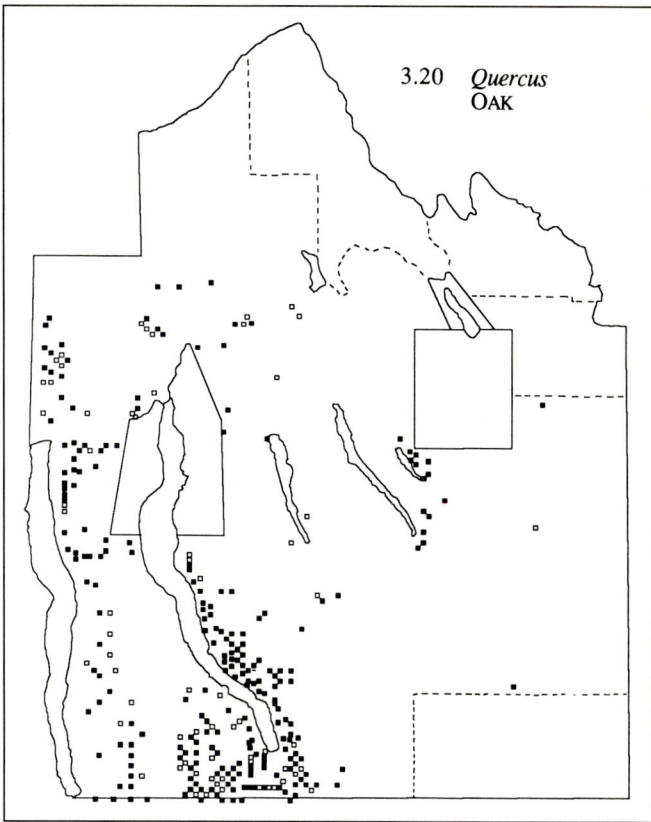
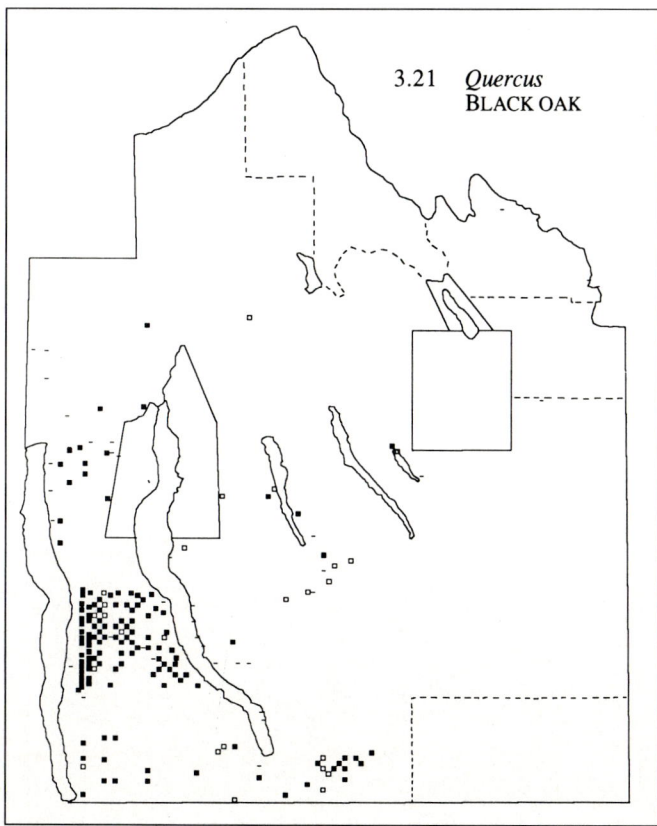
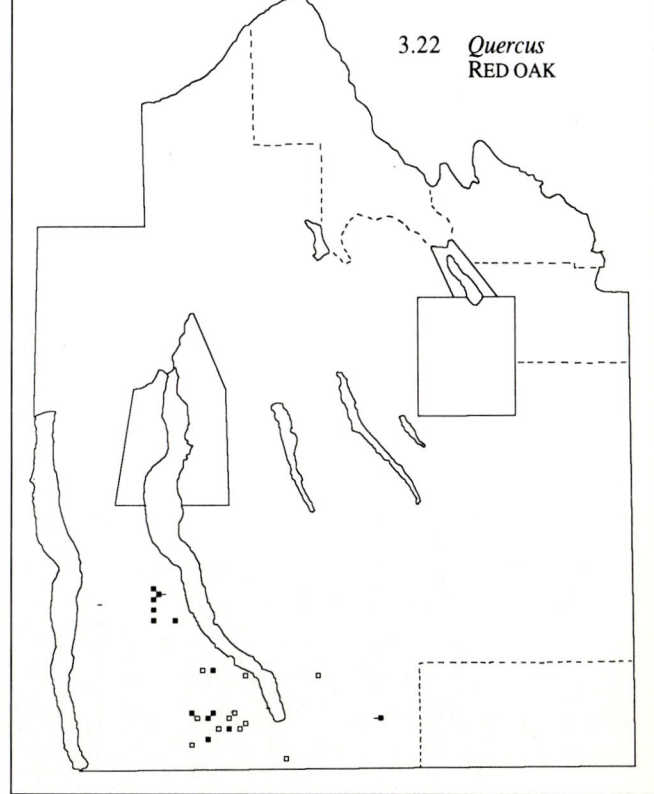

Fig. 3 (continued)

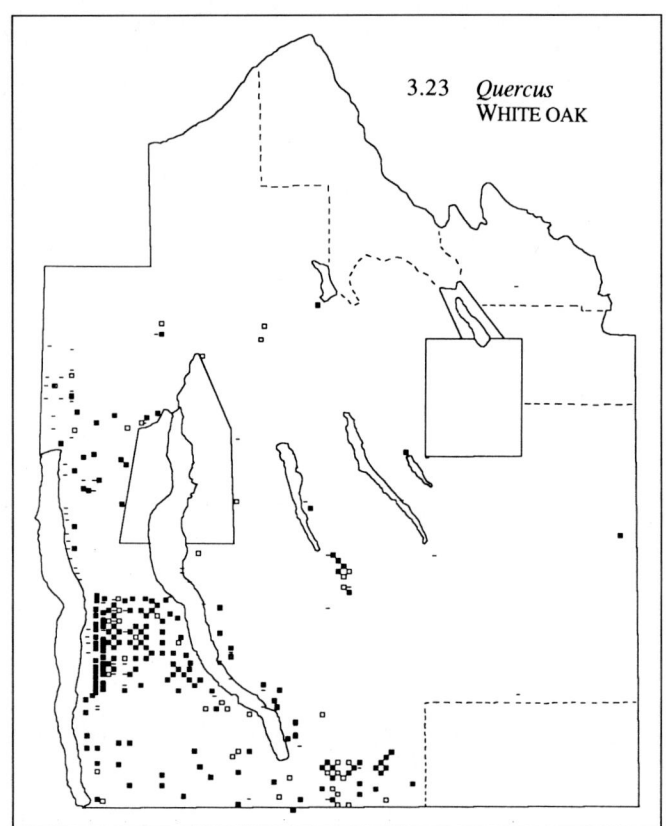

3.23 *Quercus* WHITE OAK

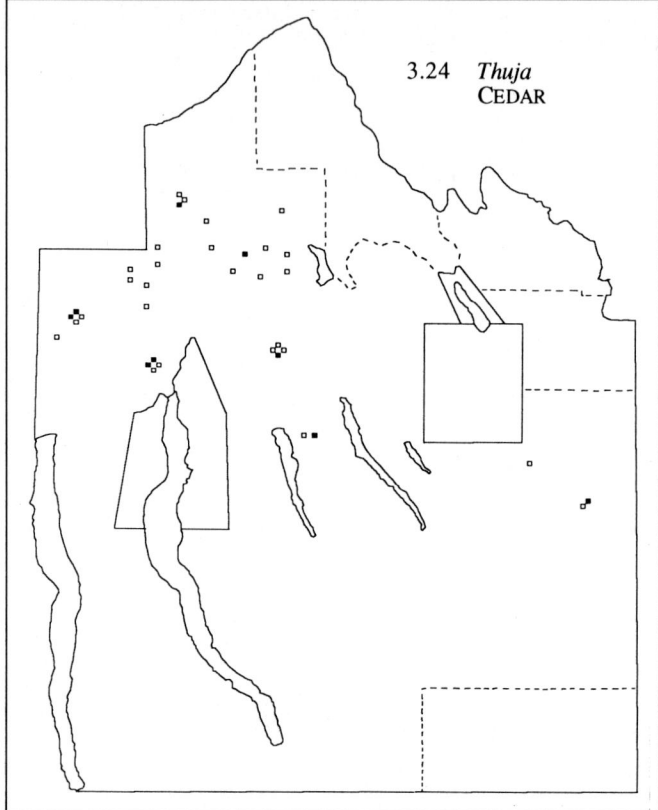

3.24 *Thuja* CEDAR

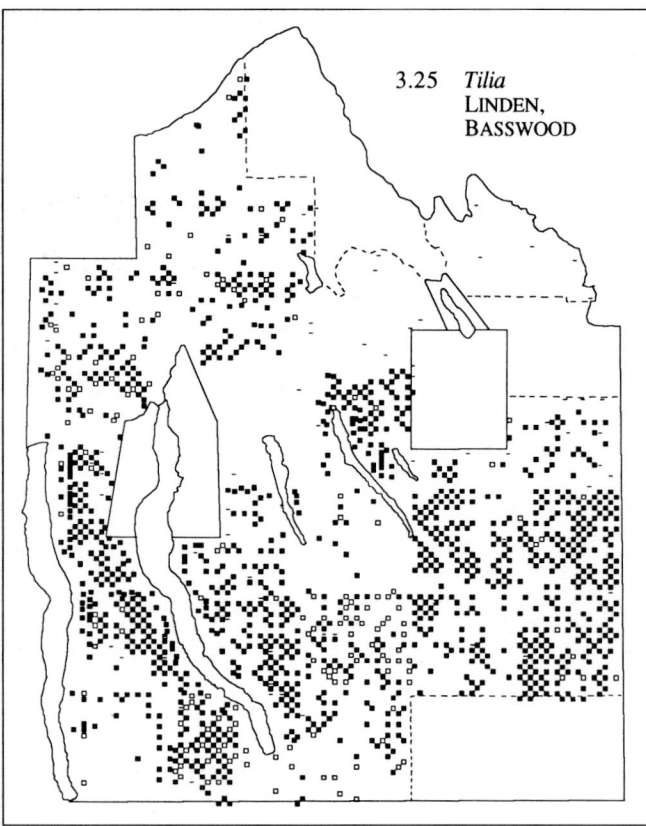

3.25 *Tilia* LINDEN, BASSWOOD

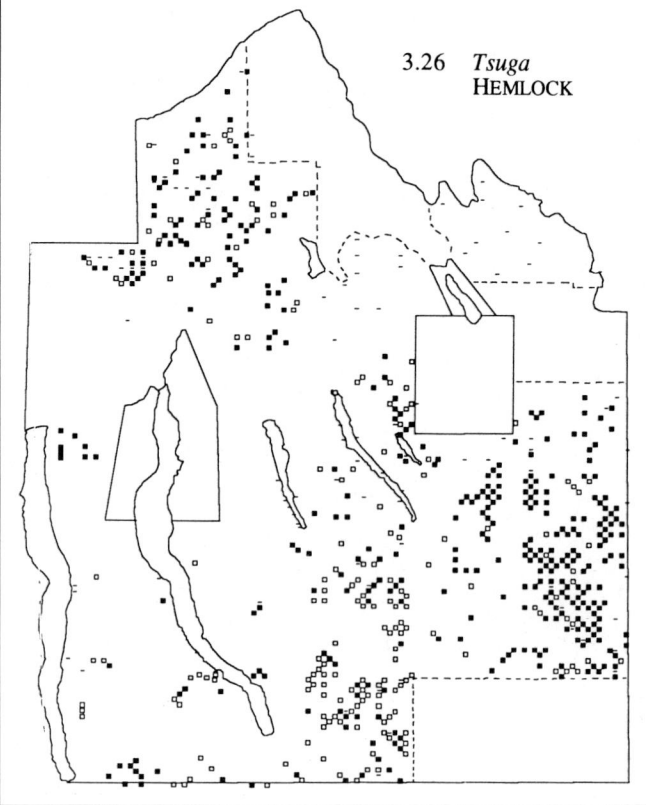

3.26 *Tsuga* HEMLOCK

Fig. 3 (continued)

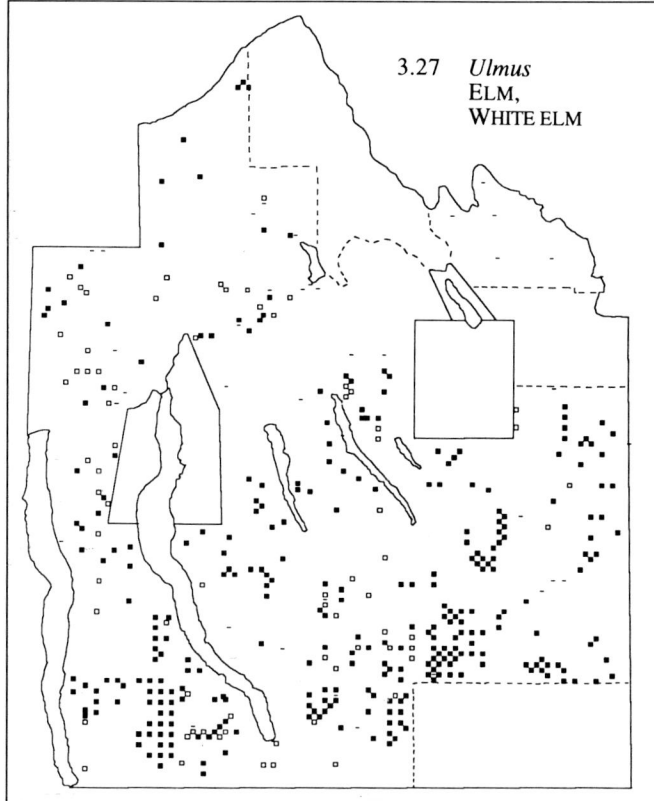

Fig. 3 (concluded).

frequency within the bounds where it co-occurred with these species.

Of the 36 species recorded at witness corners, 23 were represented as "saplings" as well as trees. A total of 15% of the witness data were recorded as "saplings." Since these occurred in all but one of the townships, there appeared to be no surveyor bias in differentiating saplings and trees. The use of saplings as witness trees was not an indication of open areas (i.e., lacking trees) since saplings were used in forested bounds; apparently these understory stems were closer than any canopy tree to the lot corner. The ranks of species as saplings or trees were similar. Of the beech, maple, hard maple, and linden at witness corners, 13% to 17% of each were saplings. Ironwood and hickory were overrepresented as saplings (>35%), and hemlock and white oak were underrepresented as saplings (<8%).

The witness stake data were primarily (92%) from 6 of the 23 townships (#14, 16, 17, 19, 20, and 22). Therefore they may reflect regional vegetation differences and surveyor bias. For instance, although beech was the most common species used for stakes (37%), 17% of the stakes were "hard" or "sugar" maple, compared to 8% for witness trees. This was because 82% of the stake data were recorded by the only two surveyors (Cuddebach and Hart) who consistently used these terms rather than "maple." The third most commonly used species was ironwood, with 10% of the stakes, compared to 1.6% as witness trees and saplings, perhaps because shade-grown *Ostrya* or *Carpinus* saplings were plentiful and the right size for stakes. It is possible that ironwood was infrequently recorded along bounds (<1%) because stems in the understory would not have been included in the boundary descriptions. Hickory, another species commonly recorded as "sapling" at witness corners, was used for 5.6% of the stakes. The two species underrepresented as witness saplings, hemlock and white oak, were uncommon as stakes. This suggests that surveyors cut stakes from species that were common in the understory, rather than selecting species that would make better stakes (e.g., be easier to cut or drive, or last longer). The most rot-resistant species readily available, chestnut, was not dramatically overrepresented in stakes compared to witness trees (1.2% of stakes vs. 0.6% of witness trees), but the sample sizes of both were small (6 stakes, 12 trees).

Regional patterns of species distribution

Maps of the occurrence of each of the common taxa, as witness trees and along lot boundaries, showed four basic patterns of spatial distribution through the Military Tract (Fig. 3). Several species were recorded throughout the region. Some were more common to the north, others to the west and south, and a few were recorded across the south or to the southeast within the Tract.

Three ubiquitous taxa were beech (Fig. 3.8), maple and hard maple (Figs. 3.1, 3.2), and linden (Fig. 3.25). These were the taxa most frequently recorded by surveyors on lot bounds (Table 3). Although not as abundant, hemlock also occurred across the tract, in the northwest and the southeast (Fig. 3.26). Hemlock

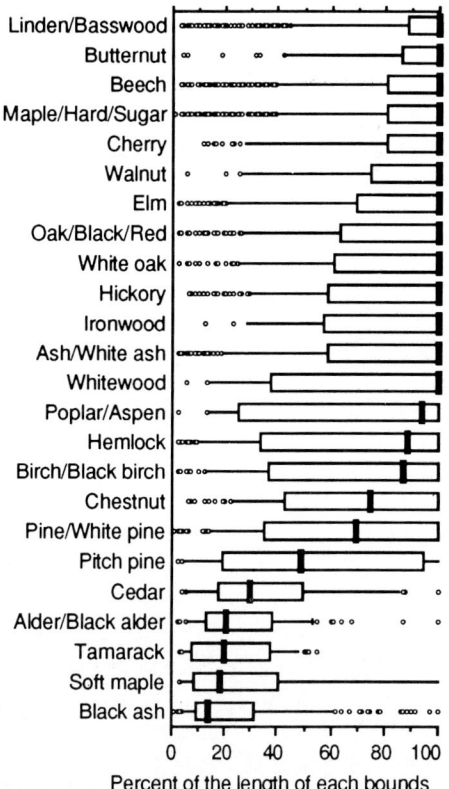

Fig. 4. Boxplots of relative bounds length for the common taxa (those recorded in >15 bounds). Relative bounds length is the distance in which a taxon was recorded as a percent of the length of that bounds. Dark bar is the median, box is 25th to 75th percentiles, horizontal lines are 10th to 25th and 75th to 90th percentiles, and dots are outliers. Bounds lengths were usually 1.5-1.7 km.

Table 4. Frequencies, sizes, and species of the different wetland types recorded by the surveyors. *Swamps* does not include "swampy" or "low" land. *Wet thickets* are "alder swamps" in which no tree species were recorded; "alder marsh" was included in the *Marsh* category. The sites described as "old beaver dams" were 2 swamps, 2 marshes, and 1 meadow.

(a) Wetland sizes, and abundance relative to the total number and distance of surveyed bounds. Maximum distance includes patches extending along >1 lot bounds. The total percent of bounds for *Wet open areas* is less than the sum because 10 bounds had two of the types.

	Percent of total number (4100)	Percent of total distance (6433 km)	Median distance (km)	Maximum distance (km)
Swamps	16.1%	3.7%	0.24	3.1
Wet open areas				
Wet thickets	1.0	0.3	0.32	1.6
Marshes	2.5	1.1	0.40	4.6
Meadows	0.2	0.03	0.15	0.4
Total	3.5%	1.4%		
Old beaver dams	0.1%	0.03%	0.38	0.8

(b) Species recorded in wetlands, relative to the number of bounds in which each category was mentioned. Since no species were named in many of the swamps and most marshes, and since species co-occurred, these do not sum to 100.

Swamps	% of 660		Marshes	% of 102
Black ash	47		Alder	7
Hemlock	9		Cedar	2
Tamarack	7		Cranberry	2
Elm	5		Grass	2
Ash	5		Tamarack	1
Cedar	5		Black alder	1
Pine, white pine	5		Black ash	1
Maple	5		Flag	1
Alder, black alder	5		Brakes	1
Soft maple	2			
Black birch	2			
White ash	1		Wet thickets	% of 42
Spruce	1		Alder	62
Birch	1		Black alder	38
Basswood	1			
Hard maple	1			
Beech	0.3			
Willow	0.3		Meadows	% of 8
Swamp oak	0.3		Grass	25
White maple	0.3		Fowl meadow grass	13
Cranberry	0.2			
Buttonwood	0.2			
Pepperidge	0.2			
Huckle-berry	0.2		Old beaver dams	% of 5
Flag	0.2		Black ash (swamp)	20
Brack	0.2			

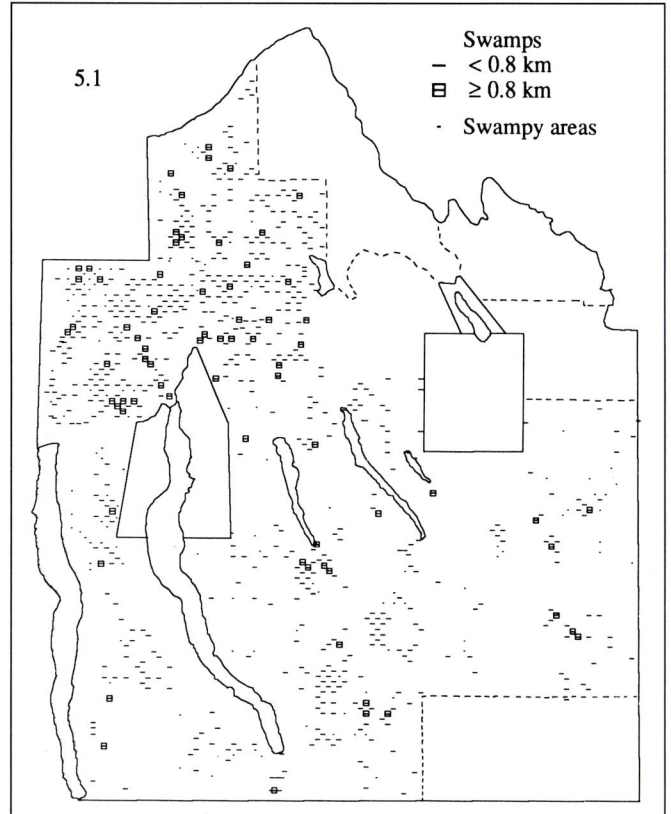
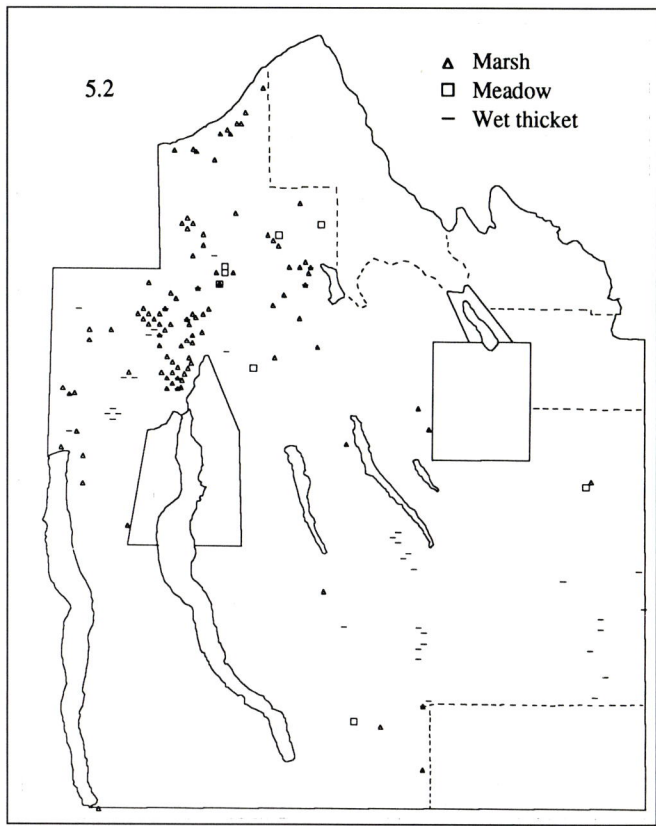
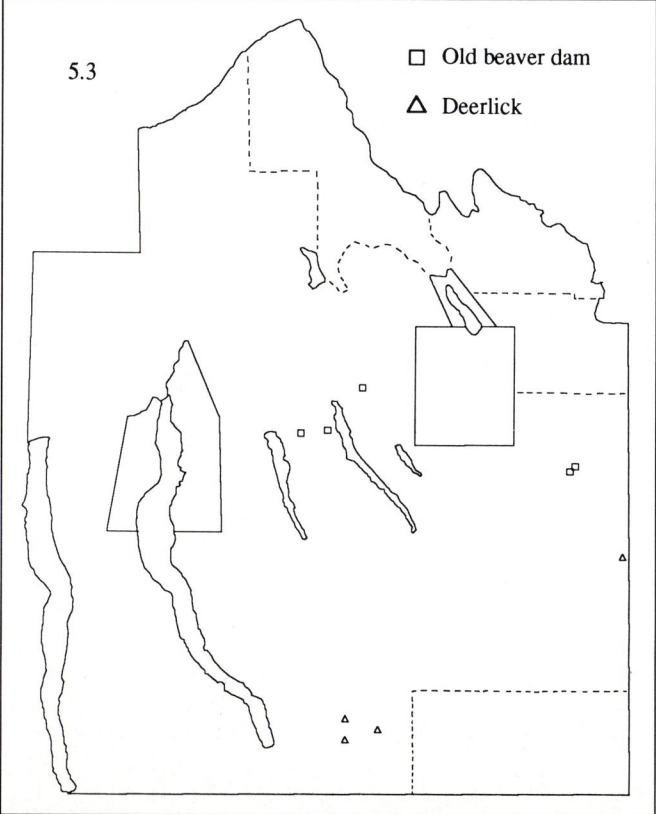

Fig. 5. Wetlands mentioned in the survey records.

    5.1) Swamps, showing small ones (<40 chains, <0.8 km) separately from large swamps, and swampy areas (e.g., "low and wet," "inclining to swamp").

    5.2) Wet open areas: marshes, meadows, and wet thickets (alder swamps with no mention of trees).

    5.3) "Old beaver dams" (2 swamps, 2 marshes, and a meadow), and "deerlicks" (apparently along creeks).

was common both as witness trees and along bounds (Table 3).

Species reported more frequently in the northern part of the study area were cedar (*Thuja*, Fig. 3.24), tamarack (Fig. 3.13), and to a lesser extent black ash (Fig. 3.10) and alder (Fig. 3.4). All of these were wetland species (Table 4), and swamps and marshes were abundant to the north (Fig. 5) in the drumlin region.

The oaks (Figs. 3.20-3.23), walnut (Fig. 3.12), and hickory (Fig. 3.6) occurred primarily in the southwest of the tract, particularly between the two large lakes, Seneca and Cayuga. On the west/southwest-facing banks of the Finger Lakes, two thirds of the witness trees were white or black oak, pine, and hickory (66% of 91), in contrast to the east/northeast-facing banks, where the most common taxa were hemlock, beech, maple, and hard maple (57% of 53). Although red oak appears to have an extremely restricted distribution within the Tract (Fig. 3.22), only three surveyors used the name, while the others presumably called it "black oak" or "oak." The surveyors recorded pine and chestnut primarily along the southern edge of the study area (Figs. 3.16, 3.7). Chestnut also occurred further to the northeast (near the Onondaga Reservation). Pitch pine (Fig. 3.17) was mentioned both in the Junius region north of Seneca Lake and southeast of Cayuga Lake, two sites where pitch pine occurs today (Seischab and Bernard 1991). Pitch pine was also recorded as witness saplings and trees on the bank of Seneca Lake. It was not recorded in any burned areas, but in one bounds in Junius the surveyor crossed beech/maple land into "open land, timber destroyed by fire" and then came "out of the burnt land into pitch pine."

Another group of species included ash, butternut, and elm (Figs. 3.9, 3.11, 3.27); these were predominantly distributed across the southern half of the tract, but also occurred in the north. Cherry (Fig. 3.19) and birch, including black birch, (Fig. 3.5) had distinctly southeastern distributions.

Some woody species occurred predominantly on the Ontario Lowland or on the Allegheny Plateau (see Fig. 1 for the boundary between the two regions). However, the abundances of each of the most common taxa (beech, maple, linden, and hemlock), relative to other species within each region, were quite similar for the Lowland and Plateau (Fig. 6). Relative to other species, black ash was much more abundant on the Lowland than the Allegheny Plateau, either based on the total bounds distance or on the number of witness trees within each region (Fig. 6). The other wetland trees were also more common on the swampy Lowland. Species with notably greater abundances on the Plateau than the Lowland were pine, pitch pine, chestnut, and the oaks (Fig. 6). Ash, butternut, and elm were relatively more abundant on the Plateau than the Lowland based on bounds distance, although the witness data did not reflect this, perhaps due to the small sample sizes for witness trees of these species.

Minor species

Many species were mentioned from 1 to 12 times in the lot boundaries and at witness corners (Table 3, Fig. 7). A number were wetland species. White maple (Fig. 3.3), like soft maple, was recorded both on the Lowland and Plateau, usually in

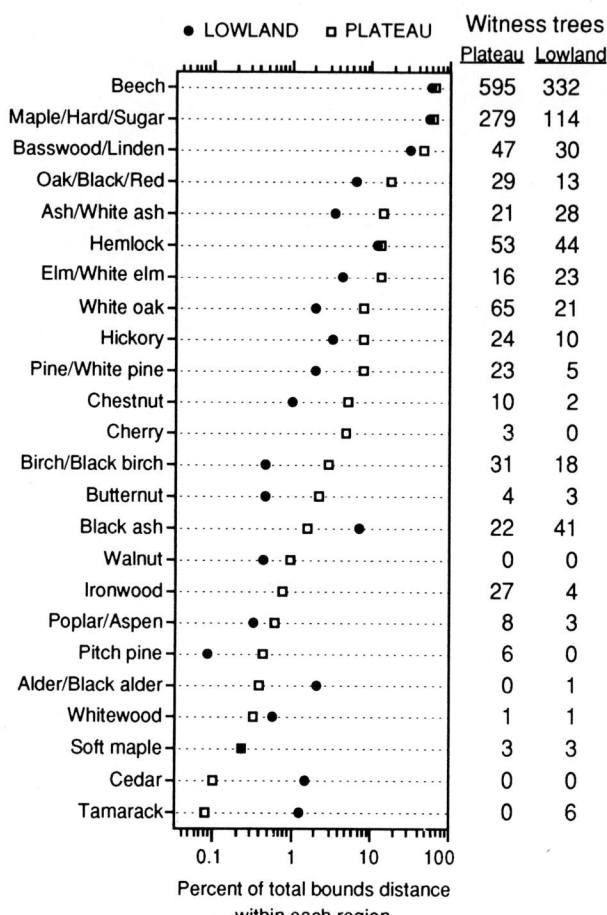

Fig. 6. Comparison between the relative abundances of the common woody taxa on the Allegheny Plateau and on the Ontario Lowland. Percent of total bounds distance for each taxon is relative to all bounds with any woody species recorded: 3323 km on the Plateau, 1192 km on the Lowland. Note that the logarithmic scale exaggerates differences for taxa with low abundance. The sample sizes for witness trees were 1284 on the Plateau (1267 were common taxa, listed here) and 708 on the Lowland (702 listed here).

swamps. Willow was mentioned only in swamps or "swampy" areas, mostly on the Ontario Lowland (Fig. 7.1). "Buttonwood" (*Platanus*) was recorded on the banks of rivers or lakes, and in swampy areas (Fig. 7.1). Spruce, which was listed exclusively in swamps and therefore was *Picea mariana*, occurred both on the Lowland and on the Plateau (Fig. 7.2). These would have been acidic bogs (Wiegand and Eames 1926); tamarack was also present in two of the swamps with spruce.

North of Seneca Lake in the township of Junius, in a region of patches of well-drained sandy soil and "a great many swampy holes," huckle-berry (Fig. 7.2) was recorded with pine and scrub oak and as underbrush in a "maple swamp." The "scrub oak land" and "scrub oak bushes" clustered in this region (Fig. 7.3) could have been *Quercus ilicifolia*. To the south, the two mentions of "scrub oak ridge" along the west facing slope at the lake may, like the nearby "scrubby oak ridge," refer not to *Quercus ilicifolia* but to stunted growth of another species. The five occurrences of "rock" or "chestnut" oak were at the extreme southern and western edges of the Military Tract (Fig. 7.3). The same distribution was seen for other components of xerophytic oak forest (e.g., pitch pine, chestnut). Swamp white oak and swamp oak (*Q. bicolor*) were confined to the Ontario Lowland (Fig. 7.3), and all five occurrences were in swamps or "low land."

Several species were associated with forest disturbances. "Thorn" (*Crataegus*), "thorns" and "briers" occurred on the Plateau (Fig. 7.4) in former clearings made by Native Americans, and in blowdowns. Currants occurred in one of the same areas of blowdown. "Thorns" also were noted in a burn: "little or no timber, occasioned I suppose by fire, but very thick covered with thorns and hazelbushes." The northern occurrence of hazel (Fig. 7.4) was as underbrush in an oak/hickory woods. Such habitats suggest *Corylus* rather than *Hamamelis* (witch hazel). Plum was recorded in two "old clearings." This was probably *Prunus nigra*, which was cultivated by the Iroquois (Hedrick 1933).

The surveyors recorded little information about herbaceous plants. In the >2700 lot boundaries with species information, herbs were mentioned only 26 times. Fourteen of these were in two townships (Solon and Dryden) where apparently the surveyors were more conscientious or perhaps better acquainted with the species. "Nettles" were associated several times with "good land;" Wyckoff (1988) commented that in the late 1790s in the Holland Land Company tract in western New York, surveyors were instructed that a rich growth of nettles was a sign of fertile land. Other species recorded in the Military Tract were mayapples, rushes, coltsfoot (more likely the native *Caltha palustris* than the European *Tussilago farfara*), maidenhair, oak of Jerusalem (perhaps the European *Chenopodium botrys*), and wintergreen. The "brack about 4 feet high" in a swamp and "large brakes" in a marsh were more likely *Osmunda* than *Pteridium*. Other wetland herbs were flag (probably *Typha*) and fowl meadow grass (*Glyceria striata* or *Poa palustris*).

Forest vegetation types

To help understand and summarize the plant communities of the central Finger Lakes region, we used the TWINSPAN computer program (Hill 1979) to classify each bounds into a community type. The TWINSPAN analysis produced eight community types (Fig. 8), each of which either had too few samples to justify subdividing further (e.g., Alder type, n = 20 bounds), or would split into overlapping categories (e.g., Pine-Oak would become Oak-Pine with some chestnut, and Pine-White oak-Chestnut).

These eight types formed three groups: swamps (Cedar/Tamarack, Black ash, and Alder), upland mesophytic forests (Hemlock-Beech, Beech-Maple-Linden, and Maple-Linden-Oak-Ash), and xerophytic woods (Oak-Hickory and Pine-Oak). The names used here for the community types are those taxa found in at least half of the bounds in the category. Since the surveyors usually recorded only 2 to 4 species in each bounds, our names for the eight types are representative of groups of species that were often listed together.

Since TWINSPAN is not based on single indicator species, and since some bounds were heterogeneous, taxa also occurred to a limited extent in other types. Therefore to describe the types more fully it is useful to mention additional species that were frequently recorded. Three swamp types accounted for 6% of the bounds in the analysis (Fig. 8). The Cedar/Tamarack type included swamps with cedar (*Thuja*) or with cedar and tamarack

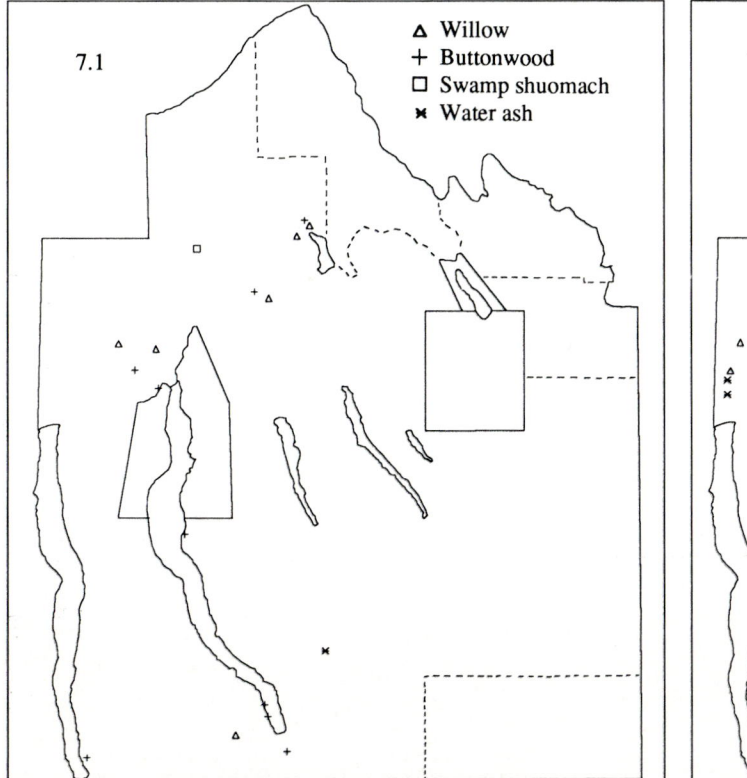
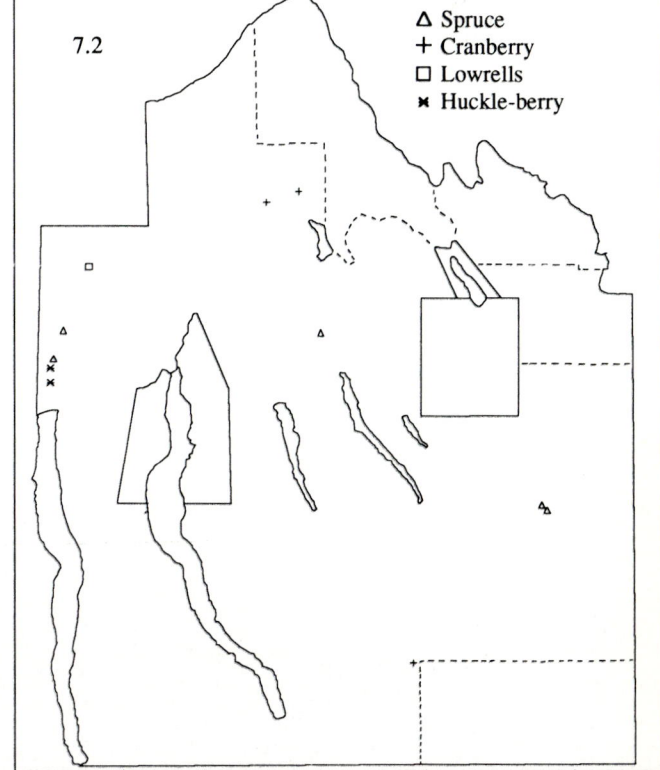

Figs. 7.1-7.6. Maps of woody taxa mentioned <15 times in the survey records. The same symbol is used for witness corners and lot bounds. Boundary information was mapped at the centers of lot bounds, so locations are approximate.

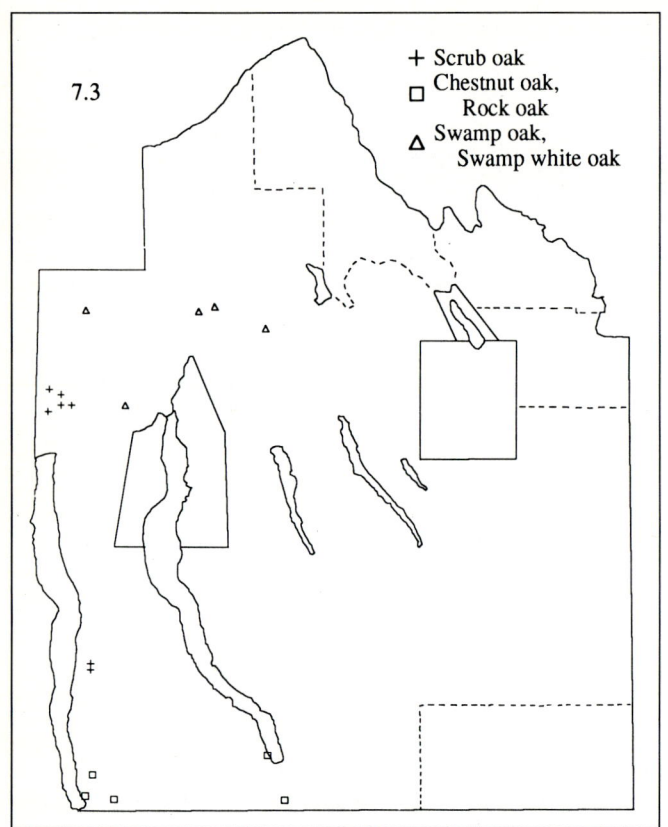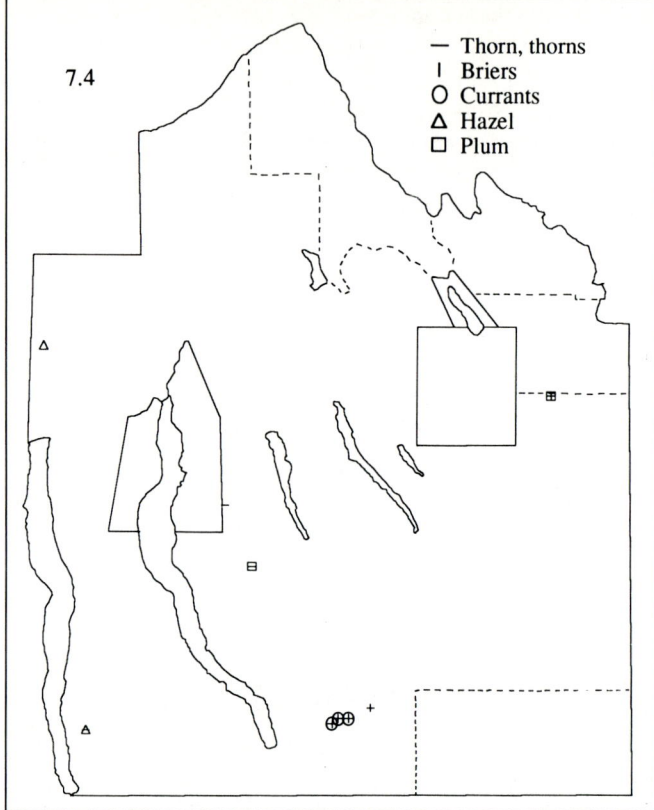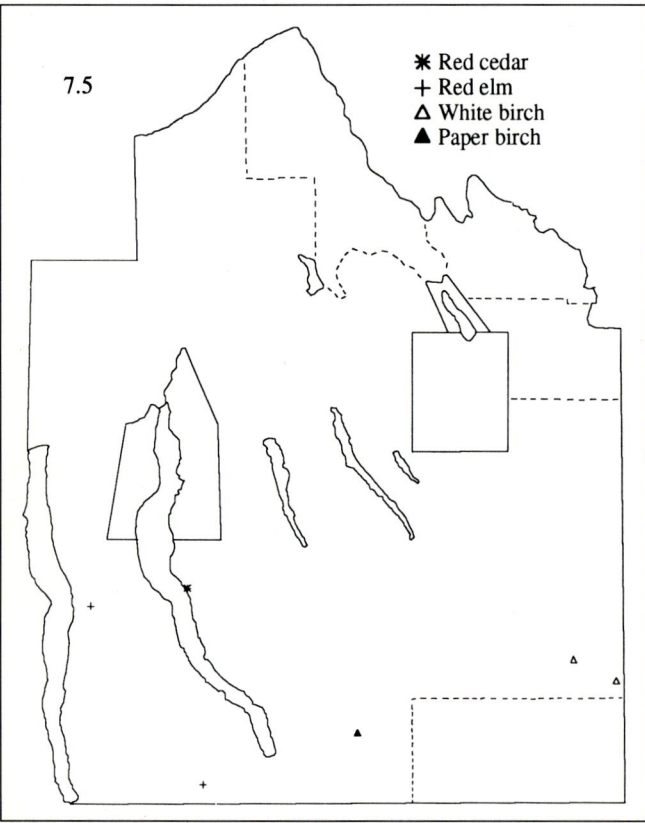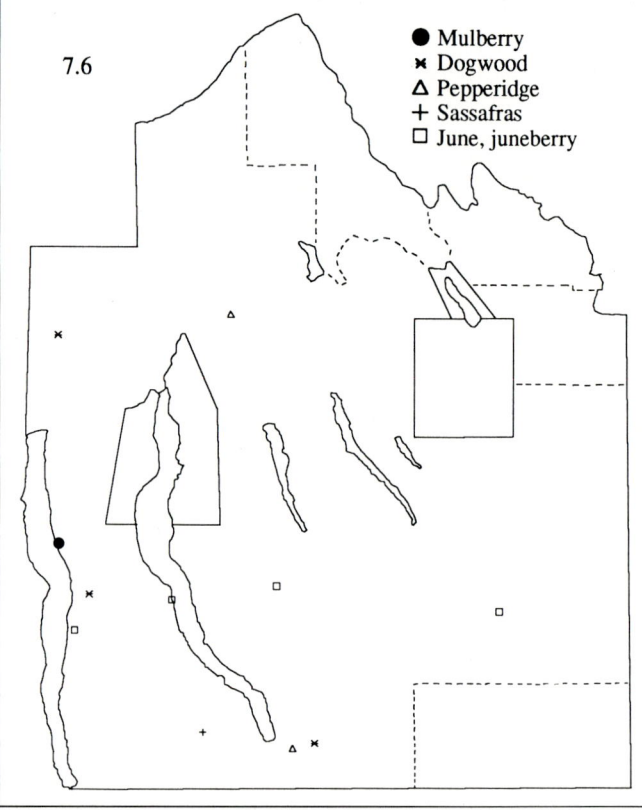

Fig. 7 (concluded).

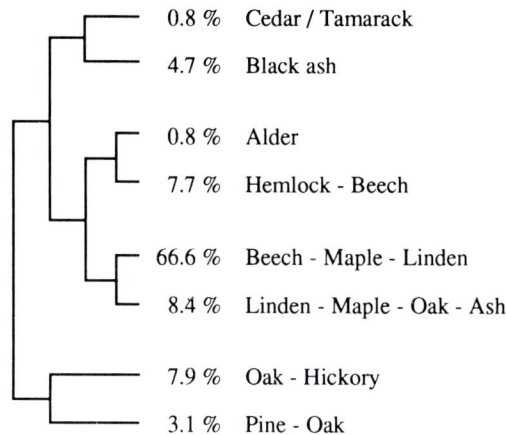

Fig. 8. Relationships among the eight community types. The TWINSPAN analysis was based on relative distance along each bounds of the 23 woody taxa mentioned >15 times. The number of bounds classified into each type is shown as a percent of the total of 2466 bounds included in the analysis. For each vegetation type, names shown are the species that occurred in at least half of the bounds in that group, in order of decreasing frequency. Tamarack was in 47% of the bounds in the first type. Oak includes "oak," "black oak," and "red oak." Alder includes "black alder." Maple includes "maple," "sugar maple," and "hard maple."

(*Larix*), which were presumably calcareous, and swamps with tamarack but not cedar, which could have been either acidic or calcareous (Wiegand and Eames 1926). The Black ash type was primarily "black ash swamp," but sometimes additional species were listed, e.g., hemlock, elm, cedar, or tamarack. Most occurrences of "alder swamp" and "black alder swamp" had no other species; a few listed elm. These same kinds of swamp also occurred on small distances within bounds classified by TWINSPAN as upland types, since we treated bounds as units. As mentioned earlier, swamps usually were short relative to bounds lengths, and although 16% of the surveyed bounds contained swamp, they totalled only 3.7% of the distance (Table 4). Swamps with black ash occurred on 11% of the bounds where surveyors recorded species (309 of 2764), but because of their small size relative to the upland forests within the same bounds, the Black ash type accounted for <5% of the bounds in the community analysis.

The overwhelming majority of bounds (83%) were in upland mesophytic forest (Fig. 8). Beech-Maple-Linden was most abundant, with 67% of the bounds. Often surveyors listed only these three taxa in a bounds, in this order, e.g., "beech, maple, linden, etc.," and "beech, sugar, and linden" (we included "hard" or "sugar" maple with "maple"). Other typical combinations recorded by surveyors in this type were "beech and maple," "maple and basswood," "beech, maple, and elm." Linden was often third or second; if the order reflected abundance, this may explain linden's low frequency as a witness tree, relative to the number of bounds in which it was listed.

The other mesophytic types, Linden-Maple-Oak-Ash and Hemlock-Beech, were less common than Beech-Maple-Linden (Fig. 8). Many bounds in the linden group had only two or three of the four taxa, e.g., "linden, maple, ash, etc.," but some did include all four: "hard maple, linden, white and black oak, white ash, and a few butternut." In some Hemlock-Beech bounds hemlock was listed first, some were "chiefly beech," and not all bounds in the type had both species.

Of the two upland xerophytic types, which accounted for 11% of the bounds, Oak-Hickory was more common than Pine-Oak. These types would also have occurred frequently in Hector (the township omitted from the TWINSPAN analysis because of scanty distance data). The Oak-Hickory type included bounds with "black and white oak, hickory, etc.," "oak, chestnut, and hickory," and the only quantitative record: "timber one-third each linden, oak, and hickory." Examples of the Pine-Oak type were "pine, oak, etc.," "white pine, white oak, chestnut," and "black and white oak, some pitch and white pine."

Geographic relations

The map of these eight vegetation types in the Military Tract shows three general regions: eastern, northwestern, and southwestern (Fig. 9). The eastern portion, including much of the central area, was predominantly Beech-Maple-Linden forest at the time of the survey. It also had much Hemlock-Beech forest, as well as a scattering of Alder thickets and Black ash swamps.

The Lowland to the northwest had much Beech-Maple-Linden forest, but it also had Cedar/Tamarack swamps and Black ash swamps (Fig. 9). Because marshes usually lacked woody plants (which were the basis for the TWINSPAN analysis), they were excluded from this figure. However, marshes also were a significant feature in this part of the study area (Fig. 5.2).

Linden-Maple-Oak-Ash, Oak-Hickory, and Pine-Oak forest types were concentrated in the southern and western parts of the Military Tract (Fig. 9). Here, Beech-Maple forest was less common than in the other two regions, though it was still represented. A narrow band of Pine-Oak, Oak-Hickory, and Linden-Maple-Oak-Ash extended northward along the western edge of the Tract.

Most of the bounds in the Cedar/Tamarack and Black ash types were on the Ontario Lowland (84% and 64%), whereas 50% of the Alder type occurred in each region. Most of the bounds in each of the five upland types (70-86%) were on the Plateau. However, in terms of relative abundance of each type within regions, Beech-Maple-Linden was equally common in both: 68% of the bounds on the Lowland, 66% on the Plateau. Since it was possible that the large Beech-Maple-Linden group might be made up of two groups, one with a species composition typical for the Lowland, and one for the Plateau, we looked at how TWINSPAN subdivided this type. Although one group had more elm, and another more hemlock, half of the Lowland bounds were in each group, supporting the conclusion that the Beech-Maple-Linden type occurred across both regions.

Species affinities

The TWINSPAN species classification produced four main groups (Fig. 10): the xerophytes, swamp species, and two groups of mesophytes. Walnut and poplar (including aspen and whitewood) were closer to the group with hickory, oaks and pines. Birch, presumably a mixture of black and yellow birch, was

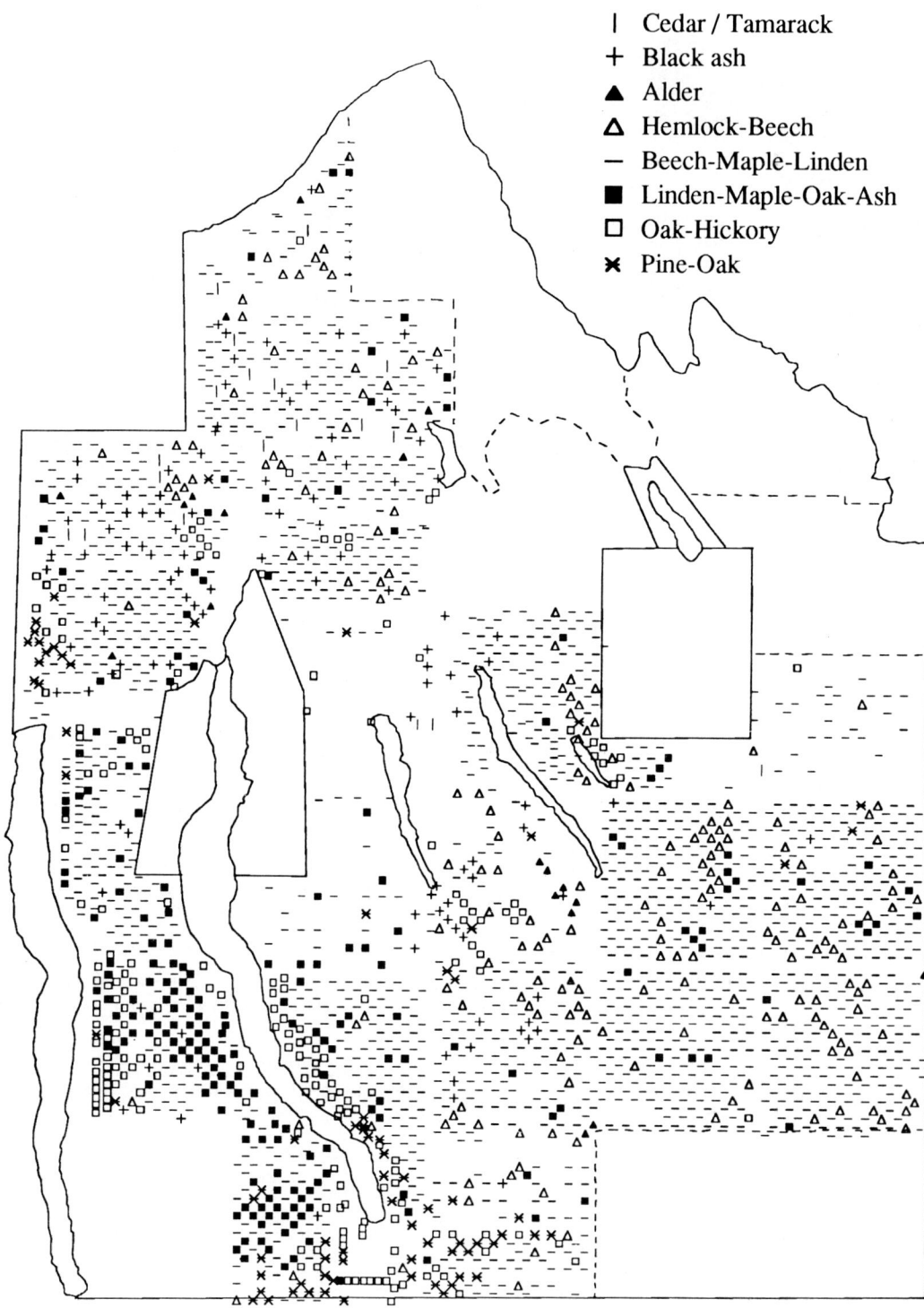

Fig. 9. Distribution of the eight vegetation types across the Military Tract.

included in the swamp group. Yellow birch occurs with hemlock and red maple in certain swamps of the region today (Mohler 1991). There seems to be little basis for distinguishing the two groups of mesophytes, in terms of ecological or environmental correlates. The grouping of ash, ironwood, and linden may reflect an association with nutrient- and base-rich sites, as Crankshaw and colleagues (1965) found in Indiana.

Quality of the land

The surveyors' descriptive terms for the quality of land or soil were often based on the vegetation (Munro 1804). The Linden-Maple-Oak-Ash type was considered best, with 77% of the bounds described as "good" or "very good" (Table 5). Bounds with Beech-Maple-Linden or Oak-Hickory were also considered to be generally good, although a few bounds in these types were

Table 5. Surveyors' terms for the quality of land or soil in each of the community types. The percent of the bounds in each type of vegetation to which these descriptive terms were applied is relative to the total number of bounds in that community type (shown in parentheses). Since not all bounds were described, and more than one term could be used (e.g., "poor, cold land"), the percents in each column do not sum to 100.

	Cedar Tamarack (19)	Black ash (115)	Alder (20)	Hemlock Beech (191)	Beech Maple Linden (1642)	Linden Maple Oak, Ash (207)	Oak Hickory (195)	Pine Oak (77)
Very good, excellent, rich, exceeding good	•	•	•	0.5%	5%	11%	5%	•
Good, fine, pretty good, middling rich	•	2%	•	6%	57%	66%	41%	14%
Middling, middling good, tolerable, tolerable good, indifferently good	•	2%	•	13%	16%	8%	19%	14%
Poor, indifferent, not very good, very poor, bad	5%	7%	•	32%	5%	•	14%	44%
Cold	•	•	•	2%	0.5%	•	•	1%
Swamp, swampy, marsh, wet, mirey, inclining to swamp	100%	100%	100%	15%	3%	8%	•	8%
Level, flat, bottom	•	•	•	4%	6%	5%	2%	3%
Rough, uneven, broken, ridge, ridgy, hilly, mountainous, hills and dales, rises and falls	•	•	10%	33%	11%	1%	5%	17%
Stony, rocky	•	•	•	2%	3%	0.5%	7%	6%

called "poor land." In contrast, Pine-Oak and Hemlock-Beech bounds were often "poor" or "indifferent." A few of the bounds with hemlock, beech, or pine had "cold" soils.

All of the bounds in the Cedar/Tamarack, Black ash, and Alder types were "swamp" (Table 5). A few of the Hemlock-Beech type were swamps, and there were also occasional swamps with maple or pine. Terms for flat, level land were occasionally used, in each of the non-swamp types. The vegetation on ridges or rough, hilly, or uneven land was often Hemlock-Beech or Pine-Oak (Table 5). The "rough" bounds in the Alder type were apparently difficult to cross rather than being hilly: "very rough and mirey along the creek," and "a very rough black alder swamp." A few bounds were noted as stony or rocky, especially in Oak-Hickory and Pine-Oak areas.

Openings in the forest

In the 1790s when the Military Tract was surveyed, the central Finger Lakes region was predominantly forested. But scattered across the landscape were patches available for species of open habitats. Some were due to natural disturbances such as wind and fire. Others were more permanent openings due to the soil being too dry and infertile to support closed forest, although many of these areas were probably also affected by fire. People, in clearing fields and setting fires, also created open areas. Beaver dams produced marsh and meadow habitat, while other marshes and meadows would have been due to topography and drainage.

Wind

Blowdowns encountered by surveyors along lot boundaries sometimes were simply noted as "entered a windfall." Others were described in more detail: "large pine trees blown up by the roots," and "blown down by a hurricane, perhaps, some ages ago, which makes it very difficult passing." All of the windfalls in the survey records were on the Allegheny Plateau, to the south within the Military Tract (Fig. 11.1). In many of the windfalls they noted thickets of saplings, including beech, birch, or cherry, and bushes such as "thorns and briers." We therefore suspected that nearby "thickets" with similar species (Table 6) were also in blowdowns. All of the windfalls and nearby thickets were in mesophytic forest, some near swamps, others on hilly uplands.

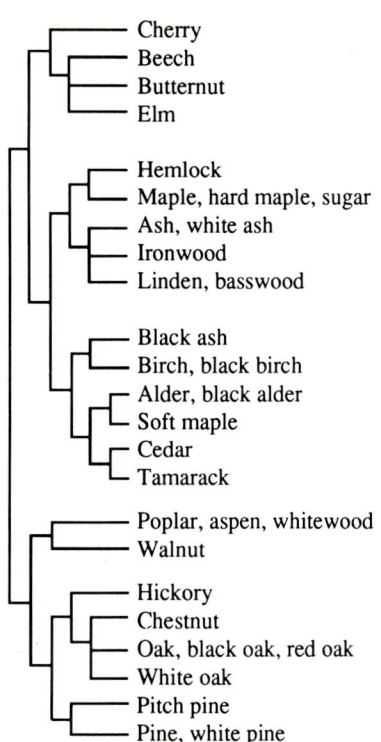

Fig. 10. A dendrogram of the relationships among 23 woody taxa, which was produced by the TWINSPAN analysis.

None of the brushy patches assumed to be blowdowns were on level land near areas recorded as "old clearings" made by Native Americans.

The median distance along bounds for windfalls and associated thickets was <0.5 km (Table 7). If blowdowns were essentially circular in shape, this distance implies a median size of 14 ha. The smallest windfall recorded was 0.2 km along a bounds, with an area of 3 ha if circular, and the smallest brushy area likely to have been a windfall ("rough land, covered very thick with young beech, birch, and maple") was 0.1 km, or 1 ha. A large blowdown at the intersection of four lots in Solon was 0.3 by 0.9 km; >20 ha. The most impressive was just north of Fall Creek in Dryden, where three areas recorded as windfall, and also a "great thicket" and a "hemlock thicket," formed a linear track that intersected the bounds of five adjacent lots. This storm track was >6 km long and 0.2 to 0.4 km across (180 ha), suggesting the path of a large tornado or thunderstorm downbursts.

Windfalls and thickets were encountered on 21 bounds, or 0.5% of the total surveyed distance (Table 7), in 9 separate patches. The sum of the lengths of windfalls and nearby thickets recorded along bounds was quite small: 0.17% of the total surveyed distance.

Fire

The surveyors noted 8 burned patches, on 10 bounds, including "timber destroyed by fire," and "has been burned over with fire which has killed all the underbrush." A few were noted as being brushy, e.g., "the timber I suppose has been destroyed by fire and is now covered very thick with underbrush." As mentioned earlier, thorns and hazel occurred in a burned area (Table 6). All of the recorded burns were on the west side of the Military Tract (Fig. 11.2), in the region dominated by oak, hickory, or pine. Several were near "scrub oak land" north of Seneca Lake in Junius, and a few were adjacent to beech-maple forest. Some fires were probably from lightning strikes, such as the chestnut ridge with "timber formerly killed with fire." Others were apparently set by people, as implied by "woods formerly burnt by firing the woods." One burn (Fig. 11.2, near the western edge of the Cayuga Reservation; see Fig. 1) was within a kilometer of a path (see Fig. 11.3). (This path led to the Iroquois village of "Canogy," shown on DeWitt's 1792 map as Connoga, west of Cayuga Lake.)

Burned areas were often larger than windfall patches, with a median distance along bounds of 0.76 km (Table 7). If these were circular, the median area of burns would be 46 ha. The longest distance recorded along bounds was 1.21 km, less than the typical size of a lot, so an unknown number of small burned areas would not have been encountered. The total distance recorded in burns was only 0.10%. However, it is possible that in addition some "open oak woods" and "open oak plains" were the result of fires, and since thick underbrush was noted in three of the burns, other brushy or "scrubby" areas in the oak/pine region may also have been fire-related.

Dry open areas

Along the western and southern sides of the Military Tract, areas of stunted growth and sparse tree canopy were common (Fig. 11.2). Patches of "scrubby timber," "thick underbrush," or "scrubby bushes," were recorded north of Seneca Lake in the sandy region in Junius, on the slopes and uplands east of Seneca Lake near a road (Fig. 11.3), and in the hills south of Cayuga Lake. Oaks and pines were common in these scrubby/brushy areas (Table 6). A large area with "scrubby beech," "black and white oak," and "thick underbrush" ran along the top of the upland between the two large lakes (Fig. 11.2).

Other areas of xerophytic or fire-related vegetation included "open oak woods," an "open barren ridge," and "but lightly timbered." White oak and "black oak" were the most common species in this type (Table 6). Most of the occurrences of open woods were on the upland and west-facing slopes along Seneca Lake (Fig. 11.2), often within a few kilometers of the Iroquois road that ran the length of the lake (Fig. 11.3). Both of the "open oak woods" just east of each of the Reservations were next to roads.

A region of "open oak plain," "beech land, open plain," and "clear oak plain" was just north of the Seneca River, between the northern ends of Seneca and Cayuga Lakes (Fig. 11.2). A road ran through this region (Fig. 11.3). This area of open land appeared to extend across at least five lots, and could have been as large as several hundred hectares.

Some scrubby or open areas were large, but like areas recorded as burns, most were <1 km along bounds (Table 7). If generally circular in shape, the median areas would be about 50 ha for open woods or scrub. Scrubby or brushy areas, open woods, and

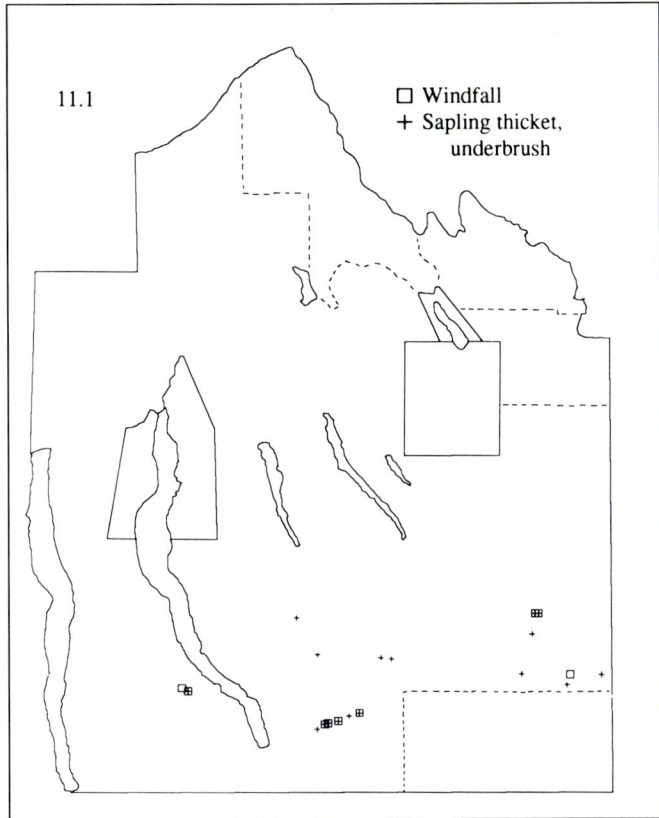
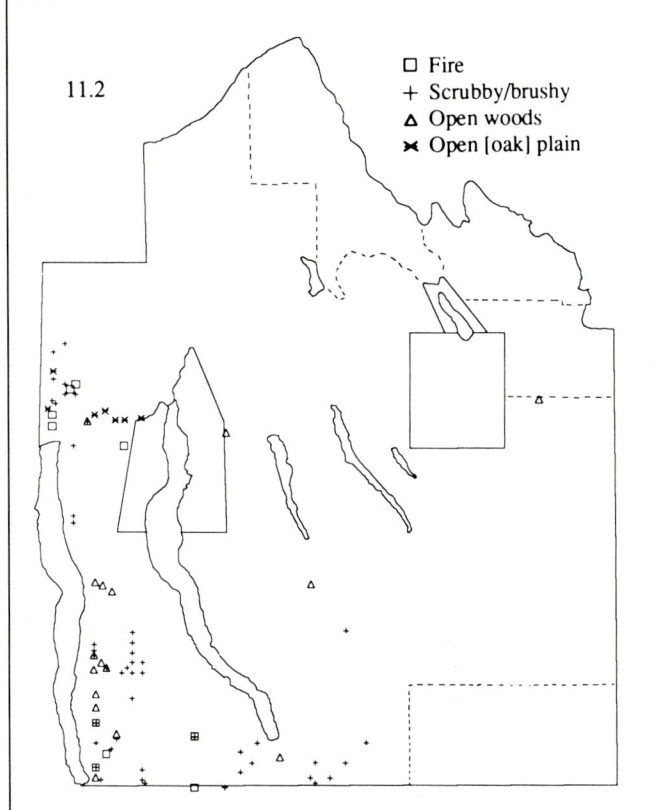
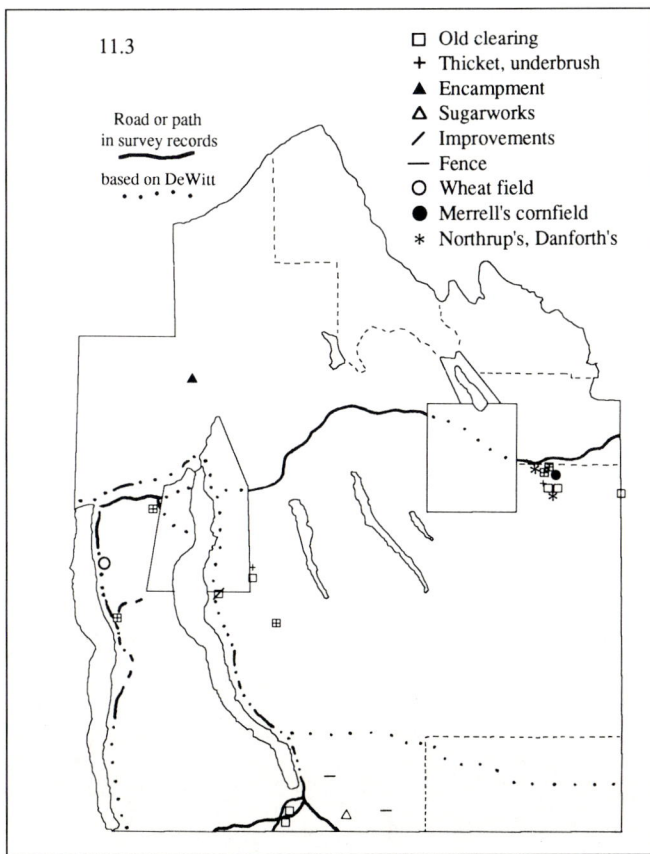

Fig. 11. Maps of disturbances and other open areas recorded by the surveyors. Locations are approximate, shown at the centers of lot bounds, except for windfalls (shown at the midpoint of their distance along the bounds, to indicate the blowdown track).

11.1) Windfalls, and nearby brushy areas likely to be windfalls. The windfalls in which the surveyors reported brush or sapling thickets are shown with both symbols.

11.2) Burned areas, "scrubby" or brushy areas in oak/pine/hickory, open woods, and open plains (usually with oak). The burns in which brush was reported are shown with both symbols (square and plus).

11.3) Areas of human activity recorded by the surveyors. These were: old clearings, two nearby brushy areas, an "Indian encampment," an "Indian sugarworks," "improvements," two mentions of "fence," a wheat field, Merrell's cornfield and meadow, and other settlers (Northrup and Danforth). Old clearings with brushy thickets are shown with both symbols (square and plus). The "improvements, lately and ancient" is shown as a slash in a square. All roads and paths recorded in the surveyors' maps and boundary descriptions are shown. Dotted portions are based on maps by Simeon DeWitt (1st sheet of his 1792 map of New York, and Map 103C in Cook 1887).

Table 6. Species recorded in disturbances and other open areas. These include saplings, bushes, trees, and burnt or uprooted trees. See text for explanation of categories. The number of bounds in which species were named and the total number of bounds are shown in parentheses, since sometimes no species were listed (e.g., "underbrush," "thicket of saplings," "scrubby timber"). Scrubby/brushy areas and open woods specifically referred to as due to fire are included in *Burned areas*, not the former categories.

SURVEYORS' CATEGORIES		ASSOCIATED BRUSHY AREAS	
	Number of bounds (# with species, of total)		Number of bounds (# with species, of total)
Windfalls	(9 of 11)	Nearby sapling thickets	(8 of 10)
Beech	7	Beech	6
Birch	6	Maple	5
Hemlock	5	Cherry, wild cherry	3
Cherry	4	Hemlock	2
Thorns	4	Elm	2
Briers	4	Linden	2
Currants	3	Birch	2
Maple	3	Poplar, popple	2
Pine, white pine	3	Black birch	1
White ash	2	White birch	1
Linden	1		
Burned areas	(3 of 10)	Scrubby/brushy areas	(32 of 45)
Thorns	1	Beech	11
Hazelbushes	1	White oak	9
Chestnut	1	Black oak	7
Oak	1	Scrub oak	7
		Pine, white pine	7
Open woods	(13 of 16)	Oak	6
White oak	7	Hickory	3
Black oak	6	Chestnut	2
Oak	5	Maple, hard maple	2
Chestnut	1	Soft maple	1
Beech	1	Hemlock	1
		Birch	1
Open plains	(7 of 7)	Pitch pine	1
Oak	4	Hazel	1
White oak	1	Huckle-berry	1
Beech	1		
Pine	1		
Old clearings	(3 of 12)	Adjacent brushy areas	(1 of 2)
Thorn, thorns	2	Thorn trees	1
Plum	2		
Briers	1		
Aspen	1		

Table 7. Disturbances and other open areas recorded by surveyors. Frequency is relative to the total number and distance of surveyed bounds. See text for explanation of the categories. The totals for percent of bounds number can be less than the sum of the types, since more than one kind could be encountered within a bounds. In calculating median and maximum distances along bounds we used the total lengths of any patches that extended along >1 bounds. The *Dry open* category does not include brush or open woods that surveyors referred to as due to fire. Distances for most *Other human activity* were not recorded. See Table 4 for sizes and numbers of bounds with *Beaver* or *Wet open* areas.

	Percent of total number (4100)	Percent of total distance (6433 km)	Median distance (km)	Maximum distance (km)
WINDFALLS	0.27	0.07	0.42	0.95
Nearby sapling thickets	0.24	0.10	0.32	1.55
Total	0.51	0.17	0.42	
BURNED AREAS	0.24	0.10	0.76	1.21
DRY OPEN AREAS				
Scrubby/brushy areas	1.10	0.67	0.84	6.40
Open woods	0.39	0.23	0.77	2.41
Open [oak] plains	0.17	0.06	0.46	1.55
Total	1.59	0.96		
FIRE + DRY OPEN AREAS	1.83	1.06		
PEOPLE				
Old clearings	0.29	0.17	0.67	2.27
Adjacent brushy areas	0.05	0.04	1.38	1.73
Subtotal	0.34	0.21	0.73	
Other human activity	0.24	[0.01–0.03]		
Total	0.54	>0.2		
DISTURBANCE:				
Wind + Fire + Beaver	0.88	0.30		
+ Human activity	1.39	>0.5		
OPEN AREAS:				
Disturbed, Dry open, Wet open	6.32	2.8		

open plains were mentioned in 1.59% of the bounds, for a total of almost 1% of the total surveyed distance. Although soils and topography would have been responsible for many of these, fire is also likely to have been involved. Combining burns and the other open areas recorded, still only 1.06% of the surveyed distance was affected.

People

By the 1790s, many of the Iroquois villages and fields in this region had been destroyed or abandoned (Peirce and Hurd 1879). The surveyors described all of the "Indian clearings" as "old" or "ancient." Since thickets of "underbrush," thorns and briers, or small trees grew in several of the old clearings (Table 6), we assumed that the two brushy areas on adjacent lots — "level land, covered with thorn trees," and "land very good, thick with small brush" — were also former clearings. A cluster of old clearings in the eastern part of the tract was near an Iroquois road (Fig. 11.3). Other clearings farther west included a "fine large flat" that was "formerly cleared by the Indians" just south of what is now Ithaca.

The lot just east of Seneca Lake that had "once been cleared by the Indians but has since grown up very thick underbrush" was shown on DeWitt's 1792 map as the location of Apple Town. In the survey records a "road from Appletown to Canogy" ran northeast out of that lot, toward an Iroquois village on Cayuga Lake. "Appletown" was an Iroquois village called Kendaia, which had been destroyed in 1779 by General Sullivan's army in his military expedition through this region (Norton 1879).

Only two apparently active Native American sites were encountered on the surveyed bounds. One was a "fine Indian encampment" in Galen (Fig. 11.3). The other was "Indian sugarworks" at the southern end of the tract, in Dryden.

In 1790-91 there were only a few settlers, at least in the region for which we had survey records. All apparently moved into areas of old clearings (Fig. 11.3). At the eastern end of the tract, William Merrell had a house, a cornfield, and a meadow. Just west of "Merrell's Improvement" was a "brook at Northrup's," and southeast of these, "Ensign Danforth's house." On the slope east of Cayuga Lake, the more recent of the "improvements, lately and ancient" was probably made by Roswell Franklin, who moved to this location (near what is now the town of Aurora) in 1783 (Hotchkin 1848). The settlement begun in 1789 by Hinepaw, Yaple, and others in what is now Ithaca (Peirce and Hurd 1879) was in "Martinus Zielie's location of 1400 acres," an area not part of the survey. "Himepough's Mill Creek" (Cascadilla Creek) was mentioned by the surveyors in Ulysses and Dryden. It was not clear whether a wheat field by Seneca Lake, and two fences southeast of Cayuga Lake (Fig. 11.3), were made by European settlers or Native Americans, since by the late 1700s the Iroquois had adopted aspects of European farming practice (W. Wykoff, pers. comm.).

Old clearings and other human activity (not including the creek) were recorded 24 times, on 22 bounds, or 0.54% of the total number. Accurate distances along bounds were not given for the camp, sugarworks, homes, fences, and two of the fields, and some bounds with cleared areas were simply described as "some old Indian clearing," or "chiefly old clearing." The median size of old clearings and nearby thickets was ±0.7 km, or >40 ha, if circular in shape. A total of approximately 0.2% of the total bounds distance was in fields and old clearings.

Wet open areas

Another type of disturbance occurred along streams; five of the bounds with meadow, marsh, or swamp were described as "old beaver dams" (Fig. 5.3, Table 4). The only one of these in which species were recorded was a "swamp, black ash — or rather an old beaver dam." There were no mentions of active beaver sites.

While beaver may have been a factor in creating other wet open areas, many marshes would have been more permanent openings created by drainage patterns. Surveyors often distinguished marshes from swamps, e.g., "out of a cedar swamp into a marsh," and mentioned tree species in only 5% of marshes, compared to >50% of swamps (Table 4). Marshes were common on the Ontario Lowland (Fig. 5.2). The largest marshes were several kilometers across, in the area now known as Montezuma Marsh. The median distance in marsh along bounds was only 0.4 km (Table 4), so they covered only 1.1% of the total boundary distance.

While many of the swamps were forested, there were patches called "alder swamp" or "black alder swamp" in which no trees were mentioned (Table 4). These wet shrub thickets were scattered across the eastern end of the Plateau, and also on the Lowland (Fig. 5.2). They may have been former beaver sites.

Beavers were also responsible for at least one of the areas of open meadow. Eight small natural meadows were mentioned in the survey records (Fig. 5.2, Table 4), including "a marsh meadow," "a clearing, apparently old beaver dams but dry," "a fine shot of clear meadow with beautiful grass," "clear meadow covered with fowl meadow grass," and "a clear fowl meadow." All were at brooks or adjacent to marshes, rather than being old clearings or fire-related.

The marshes, wet thickets, and meadows together were recorded on 3.5% of the bounds, for a total distance of 1.4% of the surveyed distance (Table 4). Although the distances recorded along bounds were generally short (medians <0.5 km), suggesting a median size of about 10 ha, this underestimates the true extent of wetlands, since from the topographic maps it was clear that marshes and swamps along stream courses or in the region of narrow north-south drumlins on the Lowland would have intersected bounds on several lots.

Estimates of total disturbed and open area

We cannot be sure that the surveyors recorded every opening encountered. However, the number of times that they clearly stated when they entered and left swamps, marshes, open woods, and thickets, noted when they crossed brooks, ridges, and roads, and commented on the slope or quality of the land even if they did not list the tree species, makes it likely that most areas large enough to intersect bounds, and sufficiently open to be noticeable, would have been recorded. It was not possible to completely separate "natural" from human impact, or "disturbance" from the more constant effects of the environment. Thus, totals reported here should be taken as approximations.

Very little of the 1790s' landscape was in patches of disturbance due to wind, fire, or beavers: only 0.3% of the surveyed distance, <1% of the number of bounds (Table 7). Old clearings were the most common of the types of human impact recorded, bringing the total area disturbed to 0.5% of the distance. Openings in wetland and dry upland areas were quite common, so altogether 6% of the number of bounds, and 2.8% of the total distance, was in open habitat in the 1790's (Table 7).

DISCUSSION

Just prior to the period of rapid settlement and widespread clearing for agriculture in the 1800s, the landscape of the Military Tract was heavily forested. Over 97% of the surveyed distance was in upland forest or wooded swamps. Within the tract, the central and eastern parts of the Allegheny Plateau were predominantly Beech-Maple-Linden and Hemlock-Beech forests. Oak-Hickory, Pine-Oak, and Linden-Maple-Oak-Ash forests dominated the southern and western parts of the Plateau, ranging north onto the Ontario Lowland at the western edge of the study area. The bulk of the Ontario Lowland was covered with a mixture of Beech-Maple-Linden and Hemlock-Beech forests, and swamp forests of black ash, cedar, or tamarack. Marshes were also common on the Ontario Lowland. The other kinds of forest openings, including windfalls, burned areas, and open oak woods, were predominantly on the Allegheny Plateau.

Species distributions within the region

Several classic vegetation maps show a difference in natural vegetation between the Allegheny Plateau and Ontario Lowland in central New York. Braun (1950), who based her vegetation regions on climate, topography and soils, included the Allegheny Plateau in the Hemlock-White pine-Northern hardwoods Region. On the Lowland, she mapped her Beech-Maple Region. Küchler's (1964) more detailed map of "potential natural vegetation" was similar for central New York except that he showed Oak forest as well as Northern hardwoods to the south. Across the portion of the Ontario Lowland for which we had survey records, Küchler's map, like Braun's, showed Beech-Maple forest.

In many ways the distinction between a Beech-Maple Region on the Lowland and Northern hardwoods on the Allegheny Plateau is subtle. Braun's Northern hardwoods and Beech-Maple Regions have many species in common; perhaps the most relevant difference is that her Northern hardwoods Region has more hemlock, white pine, and yellow birch. In the Military Tract, our Beech-Maple-Linden type accounted for two-thirds of the bounds on both the Lowland and the Plateau. At the same time, our results are consistent with Braun since our Hemlock-Beech type was common on the Plateau, as were birch and white pine. In addition, the Lowland differed from the Plateau in having many more swamp forests, especially of black ash.

For New York State in the region to the west of the Military Tract, the survey records from the Phelps & Gorham and Holland Land Company lands showed the basic pattern given by Braun: Beech/Maple forest on the lowland Till Plain, and Hemlock/White pine/Northern Hardwoods on the Plateau (Seischab 1990, and present volume). The Military Tract differed in having only modest differences in the prevalent vegetation type on the two physiographic provinces, as noted above. Such a result may be because the Ontario Lowland within the Military Tract is quite hilly (the drumlin region), whereas the Lowland (Till Plain) to the west in Seischab's study areas is much flatter (Thompson 1966). The forest types from our TWINSPAN analysis of the Military Tract survey data — upland mesophytic, xerophytic, and swamp types — were similar in many respects to the types produced by TWINSPAN in Seischab's analyses (1990, and present volume).

One of the earliest vegetation maps of New York, based on climate and topography, was that of Bray (1930), who showed sugar maple, beech, yellow birch, hemlock, and white pine on the Allegheny Plateau, and chestnut, oaks, hickories, and tulip-poplar on the Lowland. Our data do not support Bray's depiction of a chestnut/oak forest north of the Finger Lakes on the Ontario Lowland, within the region of the Military Tract. However, Bray included narrow margins along Seneca and Cayuga Lakes in this type. In the Military Tract records, oak forests did indeed occur there, but oaks, hickory, pines, and chestnut occurred in a broader band between the lakes than mapped by Bray.

A map of "primeval forests" in New York adapted from R. H. Smith (Hamilton *et al.* 1980), based on survey records and historical accounts, also showed Oak-Hickory along the west/southwest facing lake slopes, and the area of oak and hickory with other hardwoods across the upland between Seneca and Cayuga Lake. In this map, as in our results, there was no strong division between the two physiographic regions: in the area of the Military Tract a Central hardwoods type was shown extending from the Lowland onto the Plateau, with Northern hardwoods to the southeast, Hardwoods-Oak-Chestnut to the southwest, and hemlock and swamps more common to the north.

Braun (1950) described her Oak-Chestnut association extending north from the Appalachian Mountains into central New York along the Susquehanna River drainage. The more detailed information available from the survey records of the Holland Land Company (Seischab, present volume), Phelps and Gorham (Seischab 1990), and Military Tract supports Braun. Within these three areas the distribution of oaks, chestnut, and pines in the late 1700s was primarily southern, although they also occurred on coarse glacial deposits near Lake Ontario (Shanks 1966; Seischab 1990, and present volume). The distribution of oak forests in central and western New York based on survey records agrees well with Küchler's (1964) reasonably detailed map of forest types. In the Holland Land Company tract in western New York two hundred years ago, surveyors reported oak forest primarily in the south, along the Allegheny River drainage (Seischab, present volume). Coming east the next major concentrations of oak forest ranged northward across the Phelps and Gorham Purchase (Seischab 1990). The oak forests west of Seneca Lake (Seischab 1990) continued east into the adjacent western part of the Military Tract, then gradually disappeared along successive Finger Lakes.

Several factors appear to correlate with the distribution of oak forests in the Military Tract and may help explain the distribution. The first is temperature, since the oak forests in the southwestern part of the Military Tract are at the northern edge of a range that is more centrally Appalachian (Braun 1950, Küchler 1964). Oak forests within the Military Tract occurred primarily in the western half, where lower elevations, flatter topography, and the large Finger Lakes may, as Bray (1930) suggested, combine to produce a warmer climate than on the hillier plateau to the east. The second factor possibly contributing to the distribution of the oaks and related species is the continuity of habitat to the south afforded by the Susquehanna River and its many tributaries (Braun 1950). Favorable habitat for oak forest is provided by the relatively steep slopes, particularly of southern aspect, of river bluffs.

Soils and topography are also potential factors, since oak forests are often associated with xeric conditions such as gravelly or well-drained soils (Braun 1950) or ridgetops (Whitney 1991). Soils with shallow bedrock or fragipans can also be droughty, because of the restricted rooting depth (Spurr and Barnes 1980). However, the soils along the western and southern end of the Military Tract in the region where oaks and hickory were recorded were usually silt loams, not particularly xeric or shallow (USDA 1972, Cline and Marshall 1976), on the flat or moderately sloped upland between Seneca and Cayuga Lakes. This suggests that soils were not the primary factor here.

Fire would also favor oaks, which are vigorous sprouters. Earl L. Stone, Jr. (Dept. of Soil Science, University of Florida; pers. comm.) noted that in an oak forest in Cayuga County the canopy trees dated from the late 1700s, while the younger midstory and understory trees were primarily sugar maple and ash. Stone feels that the most plausible explanation for this former predominance of oak in an area that supports mesophytic forest was frequent burning by Native Americans. This is discussed further in the section on disturbance, below.

Soils

Several species showed some association with soil type, although our data were not analyzed at the level of detail necessary to make strong correlations. There was a notable hole in the distribution of black ash just south of Lake Ontario, in the only area of acidic soils north of the Allegheny Plateau (Sodus-Ira associations, Cline 1961). Similarly, only a few black ash were recorded in the southeast in the hilly region of acidic soils (Lordstown-Mardin-Volusia soils). At the scale of Cline's map (1961), cedar, i.e., *Thuja*, occurred on high lime soils.

Walnut appeared to be associated with the calcareous soils between Seneca and Cayuga Lakes (Cline 1961), but was not recorded on similar soils further east. However, walnut is at the northern edge of its range, and the population shown to the east in Little's atlas (1971) is in an area for which we did not have records, the Onondaga Reservation. Moreover, Native Americans may have influenced its distribution in this region, by planting black walnut (Wykoff 1991). Although walnut and hickory were much less abundant than oaks, the similarities in distribution suggest that climate and topography may have been as consequential as soils in determining their distribution.

At the western edge of the Military Tract approximately 10 km north of Seneca Lake is a mixture of sandy well-drained and moderately well-drained soils (Arkport-Claverack; USDA 1972), with pockets of poorly drained soils where clay underlies the sand. This was the patch where scrub oak, huckleberry, pitch pine and white pine, and swamps were recorded. The mosaic distribution of dry and wet soils, probably aided by fires in the uplands, in this case seems to have caused a corresponding mosaic of dry and wet vegetation types. Two mentions of "plain" that occurred nearby may have been on sandy soils, but because of the small-scale heterogeneity here we were not able to precisely match the surveyors' distances with a particular type on the USDA (1949) soils map. The plain with "a few scattering oak" suggests a grassy area similar to the oak openings described by Shanks (1966) in Monroe County, west of the Military Tract.

Species abundances

Beech was extremely common, as it also was further west in New York (Seischab 1990, and present volume) and in Pennsylvania (Whitney 1990). Maple, including sugar maple, and linden (basswood) were also abundant. While these taxa were recorded on a majority of the bounds, the most common witness trees were beech, maple, and hemlock. In the Phelps & Gorham and Holland Land Company tracts (Seischab 1990, and present volume), basswood had high relative frequency on survey lines but was lower in relative species weight, which was based on the order that surveyors listed co-occurring species. This supports the idea that in the Military Tract basswood was underrepresented as a witness tree, compared to its frequency of bounds, because it occurred at lower densities than the beech and maple trees in the same forests.

Based on the survey records, black ash swamps appeared to have been much more frequent 200 years ago. Swamps with black ash as the leading dominant are uncommon in the Finger Lakes region today. Mohler (1991), in his analysis of modern vegetation types in this region, did not recognize a black ash swamp type. Peattie (1966) commented that large black ash trees are now seldom seen.

There were a few native species of canopy trees present in the region (Clausen 1949) that were not referred to by the surveyors. None of these are common species. There were no mentions of cucumber tree, *Magnolia acuminata*, a species listed in the survey records from western New York (Seischab 1990, and present volume). Although it is possible some Military Tract surveyors included it in "lyn," since yellow or black linn are names for *M. acuminata* (Britton and Brown 1913), when Pursh travelled through this region in 1807, he remarked that *M. acuminata* was "very scarce about here" (Beauchamp 1923). Red pine (*Pinus resinosa*), which was not mentioned specifically (but may have been included in "pine"), occurs as natural populations in the southwestern part of the study area (Cook *et al*. 1952). Possibly "spruce" included balsam fir, *Abies balsamea*, which was not mentioned by name but is found in swamps in this area (Wiegand and Eames 1926). The likelihood that yellow birch (*Betula alleghanensis*) was included in "birch" has already been discussed.

Disturbance and other open areas

In the 1790s the major causes of natural disturbance to the forests in the Military Tract were fires and windstorms. It was not possible to determine the extent to which the fires were set by humans or lightning; both were probable causes. Fires were only recorded in the western end of the tract, primarily in the region of oak forest, on both the Ontario Lowland and the Allegheny Plateau. Areas of blowdown, on the other hand, were in mesophytic forest to the south and east on the Plateau. Disturbance was also a factor in wetlands. Beaver dams were recorded in marshes, swamps, and a meadow, and in the Finger Lakes region today, alder thickets are commonly associated with former beaver dams (P. L. Marks, pers. obs.). While most of the wetlands were primarily due to drainage and topography, the impact of beavers on the landscape would have been greater in the centuries prior to trapping for the fur trade. By the 1700s beaver populations in this region were severely depleted (Morgan 1868).

Most of the windfalls appeared to be small patches caused by thunderstorm downbursts. The blowdown 0.2 to 0.4 km wide, which intersected bounds on adjacent lots across a distance of >6 km, was potentially the linear track of a large tornado, like the recent Tionesta blowdown in Pennsylvania (Peterson and Pickett 1991). Alternatively, it may have been multiple patches of windfall from thunderstorm downbursts, which tend to be wider than tornado swaths (Canham and Loucks 1984, Fujita 1985).

In the survey records from western New York (Seischab and Orwig 1991), as in the Military Tract, windfalls were exclusively on the Plateau. Seischab and Orwig suggested that the greater topographic relief on the Plateau than the Lowland may promote greater air turbulence, and that most of the blowdowns were probably from winds associated with thunderstorms and weather fronts. Because windfalls were more abundant in the eastern tract (Phelps and Gorham) than in the western (Holland Land Company), they proposed that hurricanes coming up the Atlantic Coast might also be a factor. The trend of increasing windfall frequency did not continue eastward into the Military Tract, where many fewer blowdowns were recorded.

The percent of surveyed distance recorded in windfalls in the Military Tract was quite low, only 0.17%, especially compared to other studies in the northeast that are based on survey records (made between 1788 and 1859). For the two tracts in western New York, Seischab and Orwig (1991) reported windfalls on 1.53%, and 0.47%, of the surveyed distance on the Plateau. If these figures are converted to the entire surveyed distance (i.e., Plateau plus Till Plain), they are still higher (0.9% and 0.3%) than in the Military Tract. In hemlock-hardwood forest on the Allegheny Plateau in Pennsylvania, windfalls were recorded on 1.4% of the number of lines surveyed (Whitney 1990), compared to 0.5% of the number of bounds in the Military Tract. Windfalls were more common in lower Michigan, on 0.7% of the total distance (Whitney 1986), and even more so in Maine, on 2.6% of the distance (Lorimer 1977).

"Return time" is the time it would take for the entire distance of survey lines to be affected by disturbance, assuming the current proportion is representative. This is estimated by dividing the probable number of years that disturbances would remain visible (e.g., 15 years) by the percent of surveyed distance in disturbance at a given time. Windfall data from the Military Tract survey records give an estimated return time of about 9000 years, assuming the nearby brushy thickets were also windfalls. Return time estimates based on survey records depend both on how typical the period was during the survey and how many of the actual disturbances the surveyors recorded (Whitney 1986). Even return times of 9000 years should not be construed to indicate catastrophic windstorms are that rare. Immediately to the east of the two windfalls recorded in Ulysses in the 1790s, in 1989 a severe windstorm hit Smith Woods and nearby forests.

The quite small proportion of the survey lines reported in windfalls, with the consequently quite long return time, indicates that large windstorms had only a minor influence on the landscape of the Military Tract in 1790. Regeneration of the forest would only occasionally have occurred in openings larger than a few hectares caused by catastrophic winds. Instead, the vast majority of forest regeneration occurred in much smaller canopy gaps, which would not have been noticed or recorded by the surveyors.

It is much less clear how much of the landscape was affected by fire, since fire probably interacted with soils and vegetation, and since open woods and brushy or stunted vegetation may have burned long enough prior to the survey to not be noticeable as "burnt land." Day (1953) noted that repeated fires can result in "scrubby" oak vegetation. Seischab and Orwig (1991) reported pitch pine on 2.45% of the distance in the eastern of their two tracts and suggested that this represented fire-related vegetation. Pitch pine in the Military Tract was only mentioned once adjacent to a burn. Areas referred to specifically as due to fire were as uncommon in the survey records of the Military Tract, with 0.1% of the surveyed distance in burns, as in other regions of hardwood forest. Only one burn was reported in the two tracts in western New York (Seischab and Orwig 1991), and none in the Pennsylvania study (Whitney 1990). In contrast, conifer forests in Michigan (Whitney 1986) and Maine (Lorimer 1977) had burns recorded on 7 to 9% of the surveyed distance.

The several types of dry open habitat recorded in the Military Tract included areas likely to have been fire-related. One of these was the band of open oak plains north of the Seneca River. Although these suggest oak openings on xeric soils similar to areas described by Shanks (1966) west of the Military Tract, at least 3 of the 5 bounds with open plain near the Seneca River were on silt loams (USDA 1972), soils capable of supporting closed forest. Dudley's interpretation (1886) of historical accounts of this region was that grassy plains with scattered oak could be created by fires set by the Iroquois to drive deer, and perhaps also by clearing for fields. As a road ran through this area, and there was an Iroquois village here (Scawyace, whose cornfields were destroyed by Sullivan's army in 1779; Cook 1887), it is likely that this open land was created or maintained by repeated fires, rather than being edaphic prairie.

We suggest that fire may also have been a factor in open or scrubby oak woods. These often occurred on gentle slopes with aspects and soils that would not, by themselves, have produced the distinctive physiognomies described by the surveyors. Some of the occurrences of "open oak woods," "scrubby oak ridge," and "thick underbrush" along Seneca Lake and south of Cayuga Lake were near Iroquois roads or villages, suggesting that these may have been created by fires, which may have been used here to drive deer (Morgan 1901) or to clear brush from the road (W. Wykoff, pers. comm.). Whitney (1990) commented on the likely relationship between fires, some set by Seneca Indians, and the brushy oak and chestnut ridges along the Allegheny River in Pennsylvania. However, in the Military Tract some of the open woods, brushy or scrubby areas, and pitch pine were on ridges or steep south- or west-facing slopes, and could have been due to droughty soils.

If one assumes, arbitrarily, that half of the bounds distance in scrub, open woods, and open plains was due to fire, the estimated return time for fire would be about 2600 years. Restricting the calculation to the <15% of bounds in oak or pine forest, where most of the fires and dry open areas occurred, would produce a much shorter return time. However, such a figure would underestimate return time due to natural fires (lightning strikes), since many fires were probably set by people.

The surveyors traversed the landscape of the Military Tract at a period of transition in terms of human impact. The effects of European diseases, wars, and the Sullivan expedition had all but eliminated Iroquois populations from the region of the Military Tract (Peirce and Hurd 1879, Thompson 1966). On the other hand the rapid influx of settlers with their enormous effects on the landscape had not quite begun. Thus in 1790 the surveyors would have seen a landscape being less actively influenced by people than both earlier and later. Still, effects of people were evident, primarily from previous Iroquois activities. The introduction of European crops and domestic animals into the region had already begun by the time of the survey (Norton 1879). As early as 1807, Pursh commented on the European weeds growing along the road east of Cayuga Lake (Beauchamp 1923).

During the Late Woodland Period (ca. 1000 to 1600), Native American villages, camps, and other areas of activity were scattered both across the northern half of the central Finger Lakes region and also near Seneca and Cayuga Lakes (Hasenstab 1990). The Onondagas had large fields of corn in the 1600s in what was later the township of Pompey (Day 1953), near where "old clearings" were recorded by the surveyors. Other villages that had been to the southwest and south of Oneida Lake (Beauchamp 1905) were in areas for which we did not have survey records (Manlius and the Onondaga Reservation).

In 1779, a decade before the tract was surveyed, the Iroquois villages and crops that Sullivan's army found along the Seneca and Cayuga Lakes were destroyed, including Kendaia and its apple orchards (Norton 1879, Cook 1887). The area of "ancient" "improvements" by Cayuga Lake, near the present town of Aurora, had been the site of Chonodote, a small Cayuga village with cornfields and peach orchards (Norton 1879, Cook 1887). Coreorgonel, a village of 25 houses at the south end of Cayuga Lake (on the upland just southwest of the cleared flats mentioned in the survey records) was also destroyed (Cook 1887).

In the centuries prior to European contact, effects of Native Americans on vegetation and the landscape would have included clearing fields for corn and other crops, and cutting wood for houses and fuel (Beauchamp 1905, Day 1953). These activities would have been concentrated near villages, but areas as large as 50 ha could be affected (Day 1953). Among the woody species used for food which may have been planted near villages were black walnut and plum (Hedrick 1933, Wykoff 1991).

The Iroquois would also have had effects on the vegetation through their use of fire. Some of the eight areas noted by the Military Tract surveyors as "burnt" may have been due to Native Americans; other burns may date from the Sullivan campaign in 1779. Day (1953) and others have reviewed the potential impacts of Native Americans on the landscape, but it still remains unclear to what extent fires set by people, clearing for fields, or the cutting of wood were involved in the open oak areas. However, the Iroquois would have had more effects on the vegetation of the region than is reflected in the surveyors' scanty list of "old clearings."

CONCLUSIONS

1. *The central Finger Lakes landscape in the 1790s.* Over 97% of the distance of lot boundaries in the survey records from the Military Tract was in forest. Disturbance and wet or dry open areas were recorded on only 2.8% of the surveyed distance.

2. *Species abundances.* The predominant forest type, found throughout the region both on the Allegheny Plateau and on the Ontario Lowland, was Beech-Maple-Linden (i.e., basswood). These three were the most common taxa recorded along lot boundaries. The most common witness trees at lot corners were beech, maple (including sugar maple), and hemlock.

3. *Regional differences.* Tree species with less widespread distributions tended to occur in one of three general areas of the Tract. Swamp species such as white cedar were found predominantly in the lowlands to the north. Black cherry and birch were recorded primarily on the higher elevations to the southeast. Oak, hickory, and pine tended to occur to the south and west of the Tract, along the larger Finger Lakes.

4. *Swamps.* Forested swamps were especially abundant on the Ontario Lowland. Black ash swamps were the most common.

5. *Natural disturbance.* Relative to similar studies of survey records from other parts of northeastern U.S.A., the Military Tract had the smallest fraction of distance in windfalls, burns, or flooded by beaver dams: 0.3%. These disturbances were recorded along short distances on boundaries, often <0.8 km, but one blowdown track was >6 km long.

6. *Other openings in the forest.* Marshes were common on the Ontario Lowland. Other wet open areas were meadows and alder thickets. Open oak plains were recorded north of the Seneca River. Open, scrubby, and brushy oak or beech woods occurred on the uplands between Seneca and Cayuga Lakes and to the south.

7. *Human impact.* Former clearing by Native Americans, and settlers' homes and agricultural fields, were very sparse. Only 24 were mentioned in the >4000 bounds described. The 1790s were a period of particularly small population in the townships for which we had survey records. Effects of the Iroquois on the landscape, including oak areas possibly kept open by fire, would have been greater in earlier times.

ACKNOWLEDGMENTS

A project of this nature would not have been possible without the help and guidance of many people. Earl Stone deserves special thanks for making the first author aware that the survey records of the Military Tract existed in Albany. He also commented helpfully on an earlier version of the paper, as did Franz Seischab, William Wykoff, and two reviewers. Wykoff and Stone both informed us about the Iroquois of central New York. The staff at the New York State Archives was helpful in arranging for the use of the original surveyors' notes and maps. Beth Hedlund Marks managed to convert hours of tedious audio tapes made from the survey records into reams of accurate transcription. Katie Elliott did a masterful job of organizing the typed records into computer datafiles. David Boughton devised a way of assigning map coordinates to the thousands of computer records. Alexander Rowe and Eric Dirnbach assisted with the computer work. G. David Maddox performed the TWINSPAN analysis. Charles Mohler provided many useful comments on the interpretation of vegetation types and species distributions. We owe special thanks to Steve Gallow of Geological Sciences at Cornell for much help in the preparation of the maps. Franz Seischab's paper on the Holland Land Company tract was written about a year before ours; we thank him for his patience, allowing both papers to be published together. Finally, we thank Norton Miller and Craig Chumbley of the New York State Biological Survey for their encouragement and advice. Financial support came from McIntire-Stennis (NYC-183570), Hatch, and Mellon grants.

LITERATURE CITED

Beauchamp, W.M. 1905. A history of the New York Iroquois. New York State Education Department, Albany, New York. New York State Museum, Bulletin 78.

Beauchamp, W.M., ed. 1923. Journal of a botanical excursion in the northeastern parts of the states of Pennsylvania and New York during the year 1807, by Frederick Pursh. Dehler Press, Syracuse, New York.

Braun, E.L. 1950. Deciduous forests of eastern North America. Hafner Press, New York.

Bray, W.L. 1930. The development of the vegetation of New York State, 2nd ed. The New York State College of Forestry, Syracuse University, Technical Publication No. 29.

Britton, N.L. and A. Brown. 1913. An illustrated flora of the northern United States, Canada and the British Possessions, 2nd ed. Scribner's Sons, New York. (Republished in 1970 by Dover, New York.)

Canham, C.D. and O.L. Loucks. 1984. Catastrophic windthrow in the presettlement forests of Wisconsin. Ecology 65: 803-809.

Clausen, R.T. 1949. Checklist of the vascular plants of the Cayuga Quadrangle 42°-43° N., 76°-77° W. Memoir 291, Cornell University Agricultural Experiment Station, Ithaca, New York.

Cline, M.G. 1961. Soil association map of New York State. Department of Agronomy, Cornell University, Ithaca, New York.

Cline, M.G. 1970. Soils and soil associations of New York. Extension Bulletin 930, New York State College of Agriculture, Cornell University, Ithaca, New York.

Cline, M.G. and R.L. Marshall. 1976. General soil map of New York State. United States Department of Agriculture, Soil Conservation Service, Cornell University Agricultural Experiment Station, Ithaca, New York.

Clute, W.N. 1898. The flora of the upper Susquehanna and its tributaries. Willard N. Clute & Co., Binghamton, New York.

Cook, D.B., R.H. Smith, and E.L. Stone. 1952. The natural distribution of red pine in New York. Ecology 33: 500-512.

Cook, F. 1887. Journals of the military expedition of Major General John Sullivan against the Six Nations of Indians in 1779 with records of centennial celebrations. Knapp, Peck & Thompson, Printers, Albany, New York.

Crankshaw, W.B., S.A. Qadir, and A.A. Lindsey. 1965. Edaphic controls of tree species in presettlement Indiana. Ecology 46: 688-698.

Day, G.M. 1953. The Indian as an ecological factor in the northeastern forest. Ecology 34: 329-346.

Dudley, W.R. 1886. The Cayuga Flora, Part I. Bulletin of the Cornell University, Vol. II. Andrus & Church, Ithaca, New York.

Fenneman, N.M. 1938. Physiography of eastern United States, 1st ed. McGraw-Hill, New York.

Fernald, M.L. 1970. Gray's manual of botany, 8th ed. D. Van Nostrand Company, New York.

Fujita, T.T. 1985. The downburst. Satellite and Mesometeorology Research Project, The University of Chicago, Chicago, Illinois.

Gleason, H.A. 1952. The new Britton and Brown illustrated flora of the northeastern United States and adjacent Canada. Macmillan, New York.

Goodrich, L.L.H. 1912. Flora of Onondaga County as collected by the members of the Syracuse Botanical Club. The McDonnell Co., Syracuse, New York.

Hamilton, L., B. Askew, and A. Odell. 1980. Forest history. New York State Department of Environmental Conservation, Albany, New York State Forest Resources Assessment Report No. 1.

Hasenstab, R.J. 1990. Agriculture, warfare, and tribalization in the Iroquois homeland of New York: A G.I.S. analysis of Late Woodland settlement. Ph.D. thesis, University of Massachusetts, Amherst, Massachusetts.

Hedrick, U.P. 1933. A history of agriculture in the state of New York. New York State Agricultural Society, J.B. Lyon Company, Albany, New York.

Hill, M.O. 1979. TWINSPAN — A FORTRAN program for arranging multivariate data in an ordered two-way table by classification of the individuals and attributes. Microcomputer Power, Ithaca, New York.

Hotchkin, J.H. 1848. A history of the purchase and settlement of western New York, and the rise, progress, and present state of the Presbyterian Church in that section. M.W. Dodd, New York.

Küchler, A.W. 1964. Potential natural vegetation of the conterminous United States. American Geographical Society, New York, Special Publication No. 36.

Little, E.L., Jr. 1971. Atlas of United States trees: Volume 1. Conifers and important hardwoods. United States Department of Agriculture, Forest Service, Miscellaneous Publication 1146.

Lorimer, C.G. 1977. The presettlement forest and natural disturbance cycle of northeastern Maine. Ecology 58: 139-148.

Lutz, H.J. 1930. Original forest composition in northwestern Pennsylvania as indicated by early land survey notes. Journal of Forestry 28: 1098-1103.

Marks, P.L. and B.E. Smith. 1989. Changes in the landscape: A 200 year history of forest clearing in Tompkins County, New York. New York's Food and Life Sciences Quarterly 19: 11-14.

McIntosh, R.P. 1962. The forest cover of the Catskill Mountain region, New York, as indicated by land survey records. American Midland Naturalist 68: 409-423.

Mitchell, R.S. 1986. A checklist of New York State plants. The State Education Department, The University of the State of New York, Albany, New York. New York State Museum, Bulletin No. 458.

Mohler, C.L. 1991. Plant community types of the central Finger Lakes region of New York: A synopsis and key. Proceedings of the Rochester Academy of Science 17: 55-107.

Morgan, L.H. 1868. The American beaver and his works. J.B. Lippincott & Co., Philadelphia.

Morgan, L.H. 1901. League of the Ho-de'-no-sau-nee or Iroquois, vol. 1. Burt Franklin, New York.

Munro, R. 1804. A description of the Genesee country, in the state of New York. Pages 1169-1188 in E.B. O'Callaghan (ed.), 1849, The documentary history of the state of New York, vol. II. Weed, Parsons & Co., Albany, New York.

Norton, A.T. 1879. History of Sullivan's campaign against the Iroquois. A. Tiffany Norton, Publisher, Lima, New York.

Nyland, R.D., W.C. Zipperer, and D.B. Hill. 1986. The development of forest islands in exurban central New York State. Landscape and Urban Planning 13: 111-123.

Paine, J.A., Jr. 1865. Catalogue of plants found in Oneida County and vicinity. From the Annual Report of the Regents of the University of the State of New York, C. Wendell, Printer, Albany, New York.

Peattie, D.C. 1966. A natural history of trees of eastern and central North America, 2nd ed. Houghton Mifflin, New York.

Peirce, H.B. and D.H. Hurd. 1879. History of Tioga, Chemung, Tompkins, and Schuyler Counties, New York. Everts & Ensign, Philadelphia.

Peterson, C.J. and S.T.A. Pickett. 1991. Treefall and resprouting following catastrophic windthrow in an old-growth hemlock-hardwoods forest. Forest Ecology and Management 42: 205-217.

Sears, P.B. 1925. The natural vegetation of Ohio. I. A map of the virgin forest. Ohio Journal of Science 25: 139-149.

Seischab, F.K. 1990. Presettlement forests of the Phelps and Gorham Purchase in western New York. Bulletin of the Torrey Botanical Club 117: 27-38.

Seischab, F.K. and J.M. Bernard. 1991. Pitch pine (*Pinus rigida* Mill.) communities in central and western New York. Bulletin of the Torrey Botanical Club 118: 412-423.

Seischab, F.K. and D. Orwig. 1991. Catastrophic disturbances in the presettlement forests of western New York. Bulletin of the Torrey Botanical Club 118: 117-122.

Shanks, R.E. 1966. An ecological survey of the vegetation of Monroe County, New York. Proceedings of the Rochester Academy of Science 11: 108-252.

Sherwood, J.B. 1926. The Military Tract. The Quarterly Journal of the New York State Historical Association 7: 169-179.

Siccama, T.G. 1971. Presettlement and present forest vegetation in northern Vermont with special reference to Chittenden County. American Midland Naturalist 85: 153-172.

Smith, B.E. and P.L. Marks. Two hundred years of forest cover changes in Tompkins County, New York. Bulletin of the Torrey Botanical Club. *In press.*

Spurr, S.H. and B.V. Barnes. 1980. Forest ecology, 3rd ed. John Wiley & Sons, New York.

Thompson, J.E. 1966. Geography of New York State. Syracuse University Press, Syracuse, New York.

Torrey, J. 1843. A flora of the State of New York. Carroll & Cook, Albany, New York.

USDA. 1949. Soil survey, Ontario and Yates Counties, New York. United States Department of Agriculture, Soil Conservation Service, U.S. Government Printing Office, Washington, D.C.

USDA. 1972. Soil survey, Seneca County, New York. United States Department of Agriculture, Soil Conservation Service, U.S. Government Printing Office, Washington, D.C.

Wehle, M. 1973 (facsimile reproduction of 1866 original). New topographical atlas of Tompkins County, New York, from actual surveys especially for this atlas. Stone & Stewart, Philadelphia.

Whitney, G.G. 1986. Relation of Michigan's presettlement pine forests to substrate and disturbance history. Ecology 67: 1548-1559.

Whitney, G.G. 1990. The history and status of the hemlock-hardwood forests of the Allegheny Plateau. Journal of Ecology 78: 443-458.

Whitney, G.G. 1991. Relation of plant species to substrate, landscape position, and aspect in north central Massachusetts. Canadian Journal of Forest Research 21: 1245-1252.

Wiegand, K.M and A.J. Eames. 1926. The flora of the Cayuga Lake Basin, New York. Memoir 92, Cornell University Agricultural Experiment Station, Ithaca, New York.

Wyckoff, W. 1988. The developer's frontier: the making of the western New York landscape. Yale University Press, New Haven.

Wykoff, M.W. 1991. Black walnut on Iroquoian landscapes. Northeast Indian Quarterly 8 (2; Summer): 4-17.

Forests of the Holland Land Company in Western New York, circa 1798

Franz K. Seischab

Abstract: In 1798 forests of the Allegheny Plateau, in western New York, supported a section of the Hemlock-White Pine-Northern Hardwoods Forest. These were primarily beech-maple-hemlock communities but oak, oak-chestnut, and hemlock communities also occurred on the Plateau, principally in the southern part of the tract. The Till Plains of the Central Lowlands Province was part of the Beech-Maple Forest region. The upland forests of the Till Plains were mainly beech-maple, the bottomlands mostly black ash-elm-silver maple swamp forests. A scattering of grassland or "plains" occurred in the forest matrix of the Till Plains. Most upland mesic forests contained beech, sugar maple, or hemlock but other mesic forests of oak, basswood, and magnolia occurred. Dry-mesic forests on upper slopes contained oak, beech, and red maple. Wetland forests were of black ash-silver maple-elm, alder, or northern white cedar-larch-alder, some with black spruce.

INTRODUCTION

A considerable body of information has been gathered on the forests of the northeastern United States at the time of European settlement. In New England, Lorimer (1977) examined forests of Maine, circa 1793. The species composing greater than 10% of the witness trees in these forests were spruce, beech, balsam fir, northern white cedar, and yellow birch with hemlock also important in the southern portion of the tract examined. Whitney and Davis (1986) examined the forest history of Concord, Massachusetts from 1652 to the 20th century. They showed that the present white pine-northern red oak-red maple forest is the result of succession following various kinds of disturbance. Loeb (1987) examined witness trees and showed southeastern New York and northeastern New Jersey to have been dominated by white, red, and black oak, hickory, and chestnut. In southeastern New Jersey, oak, pine, Atlantic white cedar, and maple dominated.

In the Green Mountains of Vermont, Siccama (1971) showed beech to be the dominant on upland mid-elevation soils with spruce-fir domination at higher elevations. Further west, in the forests of the Catskill Mountains of New York, McIntosh (1962) found low and mid-elevation forests to have been beech-hemlock-sugar maple-birch.

In central-western New York, Seischab (1990) found a difference between the 1790s forest on the Till Plains of the Central Lowlands and that on the Allegheny Plateau. The Beech-Maple Forest on the Till Plains was dominated by beech, sugar maple, basswood, elm and ash whereas the Hemlock-White Pine-Northern Hardwoods Forest on the Allegheny Plateau was dominated by beech, sugar maple, hemlock, white pine, white, red, and black oak. Gordon (1940), examining the primeval forests of Cattaraugus Co. in southwestern New York, described a forest responding to a topographic-moisture gradient. He also described wetland forests of white pine-American elm and black spruce-tamarack. He found that bottomland forests were composed of cottonwood, sycamore, elm, silver maple, and black willow. He described the low elevation upland forests as beech-sugar maple and mixed mesophytic forest and oak-chestnut as being on upper slopes and ridges.

East of the Phelps and Gorham Purchase in the Military Tract, Marks and Gardescu (present volume) found that beech, maple, basswood, and oak were the most common species on the Allegheny Plateau and beech, maple, basswood and hemlock were most common on the Ontario Lowland. Black ash was much more abundant on the Lowland while the oaks were somewhat more abundant on the Plateau. They found that the wetland species, alder, white cedar, and tamarack were more abundant in the Lowland as were swamps and marshes.

Nearby on the Allegheny Plateau in northwestern Pennsylvania, Lutz (1930) and Whitney (1990) noted the occurrence of forests of beech and hemlock on mesic sites and outliers of oak forests on xerophytic sites, i.e., upper slopes, stony soils, and soils with a fragipan.

STUDY AREA

The Holland Land Company acquired, divided and sold most of the land between 78° and 79° W in New York State (Fig. 1). This 12,950 km^2 area lies between Pennsylvania and Lake Ontario and is bordered on the west by Lake Erie and the Niagara River. Beginning in 1798 the tract was divided into ranges 6.4,

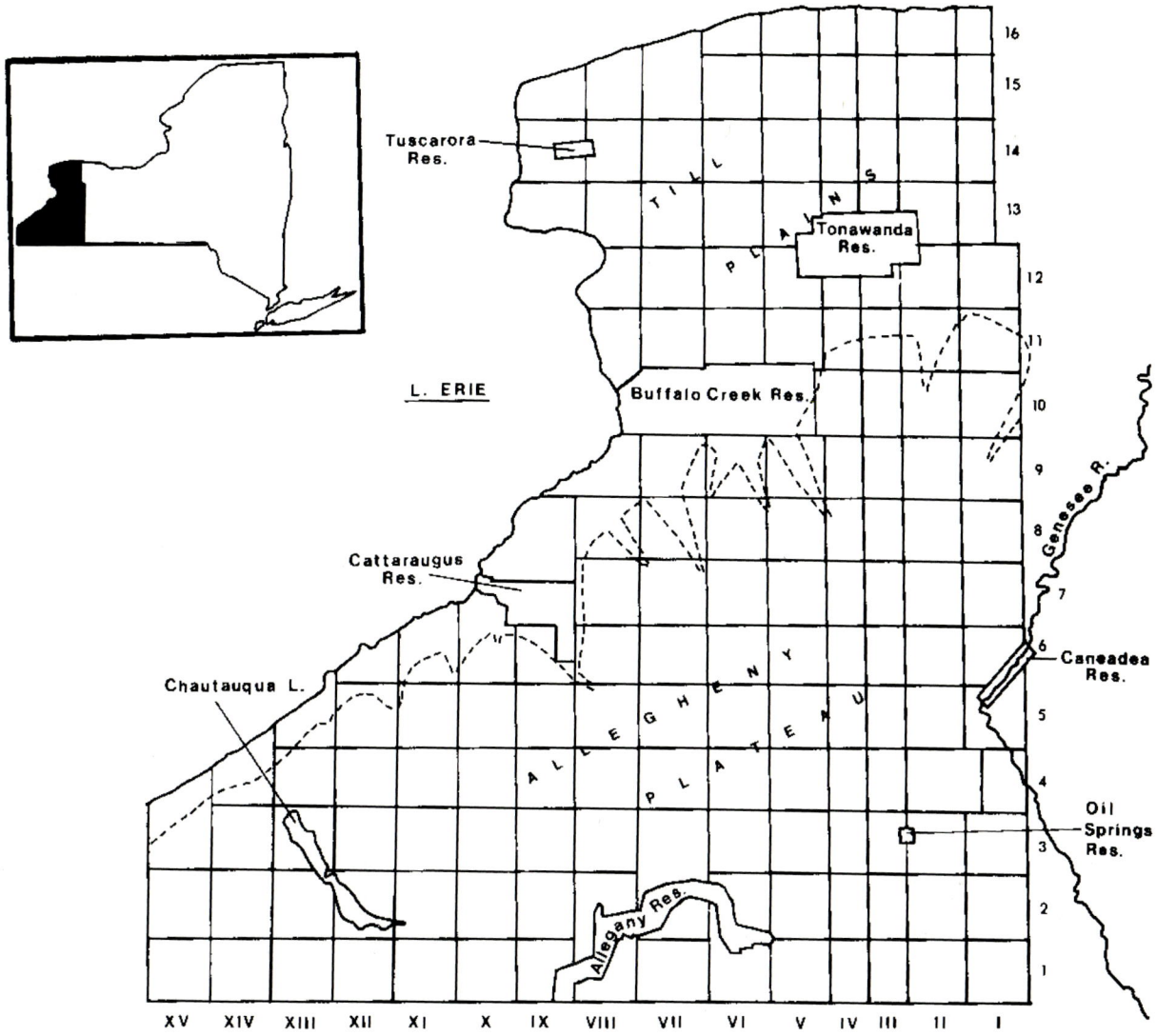

Fig. 1. A map of the Holland Company lands in western New York State. Ranges are indicated in Roman numerals. Township numbers are to the right. The dashed line represents the boundary between the Till Plains of the Central Lowlands and the Allegheny Plateau physiographic provinces.

9.6, or 11.1 km (4, 6 or 7 miles) wide and townships 9.6 km from south to north. The Buffalo Creek, Tonawanda, Cattaraugus, Allegany, Caneadea, Tuscarora, and Oil Springs Indian Reservations were established within this tract.

Agents for the Holland Land Company were more precise than previous surveyors in New York State. They used transits rather than hand held compasses and increased the accuracy of the survey by employing axemen to remove trees along the line of sight. All notes of survey were rewritten and sent to the owner bankers in Amsterdam, Holland.

These records are stored in the Municipal Archives of Amsterdam. In 1976 the Archives completed and published the *Inventory of the Archives of the Holland Land Company* (Pieterse 1976). Included in these materials is a complete set of survey notes of the Holland Company lands. The above records are on microfilm at the Holland Land Company Project at the State University of New York College at Fredonia. A description of this project can be found in Safran (1988).

The purpose of this study was to: 1) use the survey notes to reconstruct the vegetation of the 1798 forests of the Holland Land Company; and 2) compare and contrast the results with earlier studies in adjacent areas.

METHODS

Survey notes provide two kinds of vegetation information: bearing tree data and line descriptions. In the Holland Company survey, information on two bearing trees was provided at the end of each mile and four trees at each of the four corners of each

Table 1. A. Initial species weight of the leading species in the survey notes list. B. Species weights based on the niche preemption hypothesis as described in the text. If a sample contains four species then the first species has a value of 40, the second 24, the third 14.4 and the fourth 8.6. C. Relative species weights.

A.

	Number of Species								
	1	2	3	4	5	6	7	8	9
Initial Weight	100.0	66.7	50.0	40.0	33.3	28.6	25.0	22.2	20.0

B.

	Number of Species								
	1	2	3	4	5	6	7	8	9
Species Weights	100.0	66.7	50.0	40.0	33.3	28.6	25.0	22.2	20.0
		22.2	25.0	24.0	22.2	20.4	18.8	17.3	16.0
			12.5	14.4	14.8	14.6	14.1	13.4	12.8
				8.6	9.9	10.4	10.6	10.5	10.2
					6.6	7.4	7.9	8.1	8.2
						5.3	5.9	6.3	6.6
							4.5	4.9	5.2
								3.8	4.2
									3.4

C.

	Number of Species								
	1	2	3	4	5	6	7	8	9
Relativized Species Weights	100.0	74.8	57.3	46.0	38.4	33.3	28.8	25.6	23.1
		25.2	28.6	27.6	25.6	23.5	21.6	20.0	18.5
			14.3	16.6	17.1	16.8	16.2	15.5	14.8
				9.9	11.4	12.0	12.2	12.1	11.8
					7.6	8.5	9.1	9.4	9.5
						6.1	6.8	7.3	7.6
							5.1	5.7	6.1
								4.4	4.8
									3.9

township. In a 6 X 6 mile (9.6 X 9.6 km) township, this would be a sample of 56 trees for a 93 km² area or 0.6 tree/km², a rather sparse sample. However, line descriptions provide tree species lists for several segments of each surveyed mile. Thus a larger number of species, representing a greater number of individuals, is provided for each surveyed mile by line descriptions than by bearing tree descriptions. These are the most important data used in reconstructing the forests at the time of the original surveys.

I assumed that species recorded in the line descriptions were arranged in order of their importance or abundance. This is reasonable, since humans tend to itemize objects in descending order of size, number, volume, or importance. By 1804 Federal surveyors in the Midwest were instructed to list tree species in this manner (White 1984).

If species are listed in order of importance they can then be quantified to approximate that importance. This is because the 5 or 6 most important tree species in a community invariably approximate a linear dominance-diversity relationship. This was shown by Whittaker (1975) for a species poor subalpine forest and for trees in the Brookhaven oak-pine forest (Whittaker 1969). Bazzaz (1975) reported on a forty year old successional field in which the most important species formed a straight line dominance-diversity curve. Linear relationships are expected in "small samples from communities for which the curves are sigmoid" (Whittaker 1965) and since species lists from survey notes represent such "small samples" they should exhibit such relationships. Therefore, I assumed that the communities in the survey notes formed straight line dominance-diversity curves.

Each surveyed mile was treated as a linear plot and will be referred to as a sample. Relative species weights (RSWs) for each species in each mile were calculated as follows. Species were weighted from last to first, in numerical order from 1 to i, as listed by the surveyor. The relative weight of the first species was initially approximated as that species abundance (Table 1A). First approximations for the remaining species in the list were determined based on the fundamental supposition of the niche-preemption hypothesis, namely, that each successive species allocates the same proportion of the remaining resources as the most important species allocated for itself (Table 1B). Final approximations were determined by relativizing the first approximations (Table 1C). These relative species weights form straight line dominance-diversity curves as described earlier.

The RSWs, in Table 1C, were used to generate a species by mile data matrix of RSWs based on species lists in the survey notes. For example, if a surveyor listed a mile as containing beech, sugar maple, basswood, and white ash, these species were assigned RSWs of 46.0, 27.6, 16.6, and 9.9 respectively (Table

1C). In another mile a surveyor may have listed the above species in the first half mile and white oak, black oak, and hickory in the second half. In this case beech, sugar maple, basswood, and white ash would be assigned RSWs of half the values seen in column 4, Table 1C, or 23.0, 13.8, 8.3, and 4.9 respectively. White oak, black oak, and hickory would be assigned half the values in column 3, Table 1C, or 28.6, 14.3, and 7.1 respectively. The sum of the RSWs for the mile would be 100.

A species by mile matrix of RSWs was generated as described above and subjected to the classification program TWINSPAN (Hill 1979a), a polythetic divisive technique designed to identify subgroups or communities within large data matrices. Detrended correspondence analysis (DCA) (Hill 1979b) was used in order to detect ecological trends and relationships in the matrix. Relative frequency was determined as the percent of surveyed miles in which a species was recorded in the survey notes. Relative frequency data from the Holland Company survey, the Phelps and Gorham survey (Seischab 1990), and from the Military Purchase (Marks and Gardescu, present volume) were relativized to 100%. These were compared to the witness tree frequency data in the Catskill (McIntosh 1962) and Pennsylvania studies (Whitney 1990, Lutz 1930) using Sorenson's (1948) similarity coefficient: $C = 2W/A + B$, where A and B are the frequencies of all the species found in each of two communities to be compared and W is the sum of the lesser values for the species common to the two stands.

Maps of the relative species weights were generated in order to determine the regional distribution of each of the species.

Vascular plant nomenclature follows Mitchell (1986).

RESULTS

Relative Frequency

Common names of species included in the survey were usually the same as those used today (Table 2). Assumptions were made in the interpretation of some of the common names used by surveyors (Table 2).

The frequency and RSW data sets were divided according to the physiographic provinces in which the samples occurred and are listed in Table 3. Beech (*Fagus grandifolia*) and sugar maple (*Acer saccharum*) were the two most widely distributed species with approximately the same relative frequencies on both the Till Plains and the Allegheny Plateau. Mesic and wet mesic species such as elm (*Ulmus*), basswood (*Tilia americana*), white ash (*Fraxinus americana*) and black ash (*F. nigra*) all had greater frequencies on the Till Plains than on the Allegheny Plateau. Those species with greater frequencies on the Plateau than the Plains were hemlock (*Tsuga canadensis*), white pine (*Pinus strobus*), chestnut (*Castanea dentata*), yellow birch (*Betula alleghanensis*), and *Magnolia acuminata*. Most species were recorded on both the Allegheny Plateau and the Till Plains. *Quercus bicolor*, *Picea mariana*, *Zanthoxylum americanum*, *Kalmia* sp., and *Abies balsamea* were reported only from the Plateau while *Acer negundo*, *Liriodendron tulipifera*, *Thuja*

Table 2. List of species encountered in the survey notes for the Holland Company Lands.

Species	Notes of Survey Descriptions
Abies balsamea	fir
Acer negundo	box elder
Acer rubrum	maple, soft maple[1]
Acer saccharinum	maple, soft maple[1]
Acer saccharum	maple, sugar maple
Alnus incana	alder
Betula lenta	black birch
Betula alleghanensis	birch[2]
Carpinus caroliniana	hornbeam
Carya spp.	hickory
Castanea dentata	chestnut, chesnut
Cornus florida	dogwood
Corylus spp.	hazel
Crataegus spp.	thorn
Fagus grandifolia	beech
Fraxinus americana	ash, white ash
Fraxinus nigra	ash, black ash
Juglans cinerea	butternut
Juglans nigra	walnut, black walnut
Kalmia spp.	laurel
Larix laricina	tamarack
Liriodendron tulipifera	whitewood, tulip tree
Magnolia acuminata	cucumber tree, cucumber
Nyssa sylvatica	pepperidge
Ostrya virginiana	ironwood
Picea mariana	black spruce, spruce
Pinus strobus	pine, white pine
Platanus occidentalis	buttonwood, sycamore
Populus spp.	poplar, aspen, aspine
Prunus serotina	cherry
Quercus alba	white oak, oak[3]
Quercus bicolor	swamp white oak, swamp w oak
Quercus montana	rock oak
Quercus palustris	swamp oak
Quercus rubra	red oak
Quercus velutina	black oak
Salix nigra	willow
Taxus canadensis	shinwood, shin[4]
Thuja occidentalis	cedar[5]
Tilia americana	basswood, lyndon
Tsuga canadensis	hemlock
Ulmus spp.	elm[6]
Vitis riparia	grape, wild grape
Zanthoxylum americanum	prickly ash
Other Designations	
Agricultural fields	fields, cleared
Marsh	marsh, mire, bog
Plains	plains
Rock outcrops	rock bottom
Shrubs	shrubs, brush

[1] Often there was no distinction made between *Acer saccharum* and *A. rubrum* when "maple" was used in the notes of survey. For analytic purposes "maple" was interpreted to mean sugar maple when found in association with beech, basswood, and other mesic species. It was interpreted to have been red maple when found in association with oaks and hickories. When the term "maple" was used in wetland situations in association with elm and black ash it was interpreted as being *A. saccharinum* and as *A. rubrum* when in association with cedar or larch.
[2] Some of the individuals referred to as "birch" may have been *B. lenta*. Although some of the surveyors differentiated between *B. lenta* and *B. alleghanensis* it is not clear that all surveyors made this distinction.
[3] The term "oak" was often used in this survey, an obvious reference to mixed oak stands. It is presumed that these stands contained both white and black or red oaks. For analytical purposes, where "oak" was recorded in the notes of survey both white and either black or red oak was quantified in the data set.
[4] "Shinwood", which appears in the notes of survey, has a single reference to its analogue (Britton and Brown 1913), *Taxus canadensis*, which was brought to my attention by Gardescu (personal communication). Siccama (personal communication) believes this may also be *Amelanchier*, used as a cathartic and referred to as "Shittum wood" in Vermont surveys. From the associated descriptions in the survey records this term was being used to describe a type of shrub and so was most likely *Taxus*.
[5] The notes of survey use the term "cedar", here interpreted as being *Thuja occidentalis* not *Juniperus virginiana* since all the "cedar" references occurred in wetlands in association with tamarack or black ash.
[6] "Elm" was used in the survey to represent both *Ulmus americana* and *U. rubra* since it occurred in both upland and bottomland forests.

Table 3. Relative frequency and relative species weights for the Holland Company Lands, for the Allegheny Plateau and Till Plains portions of the tract. There were a total of 2049 surveyed miles, 700 on the Till Plains and 1349 on the Plateau.

	Relative Frequency			Relative Species Weight		
	Total	Allegheny Plateau	Till Plains	Total	Allegheny Plateau	Till Plains
Fagus grandifolia	92.5	92.1	93.1	22.4	23.4	20.5
Acer saccharum	83.6	84.5	81.9	20.3	22.6	15.6
Tilia americana	69.0	61.1	84.7	8.2	7.1	10.5
Tsuga canadensis	56.9	64.5	41.9	10.0	11.4	7.3
Ulmus spp.	52.9	43.4	71.6	6.2	4.7	9.1
Fraxinus americana	45.7	33.4	69.7	3.8	2.6	6.3
Quercus alba	30.5	25.7	39.7	3.8	3.6	4.2
Betula alleghanensis	29.9	39.2	11.7	2.7	3.7	0.8
Pinus strobus	25.6	34.2	8.7	3.4	4.7	0.9
Fraxinus nigra	24.1	13.3	45.3	3.6	1.7	7.3
Quercus velutina	20.8	18.3	25.9	2.4	2.6	1.9
Castanea dentata	20.4	25.0	11.4	2.4	3.1	0.9
Acer rubrum	20.1	19.4	21.4	2.2	2.1	2.5
Carya spp.	19.5	7.6	43.0	1.4	0.6	2.9
Magnolia acuminata	17.3	22.8	6.6	1.3	1.8	0.4
Prunus serotina	11.0	16.5	0.3	2.9	1.1	6.4
Populus spp.	9.3	3.9	20.0	0.6	0.2	1.4
Ostrya virginiana	8.5	5.7	14.0	0.5	0.4	0.7
Juglans cinerea	6.7	6.3	7.6	0.5	0.5	0.4
Acer saccharinum	5.6	2.6	11.6	0.6	0.3	1.0
Quercus rubra	4.8	5.2	4.1	0.4	0.5	0.3
Alnus incana	4.2	3.5	5.6	0.4	0.3	0.7
Juglans nigra	3.0	0.7	7.4	0.3	0.1	0.7
Platanus occidentalis	1.9	2.6	0.5	2.8	0.2	8.0
Larix laricina	1.6	0.3	4.1	0.2	<0.1	0.6
Thuja occidentalis	1.5		4.6	0.2		0.7
Salix nigra	1.5	0.3	3.9	0.2	<0.1	0.4
Crataegus spp.	1.4	2.1	0.1	0.1	0.1	<0.1
Liriodendron tulipifera	1.0		2.9	0.1		0.4
Quercus montana	0.9	1.2	0.3	0.1	0.2	<0.1
Quercus palustris	0.9	0.3	2.0	<0.1	<0.1	0.1
Abies balsamea	0.9	1.4		0.1	0.1	
Betula lenta	0.5	0.4	0.6	<0.1	<0.1	<0.1
Shrubs	0.4	0.1	1.1	<0.1	<0.1	0.1
Marsh	0.4	0.2	0.7	0.1	0.1	0.2
Plains	0.3	0.1	0.9	0.1	<0.1	0.3
Kalmia spp.	0.3	0.4		<0.1	0.1	
Carpinus caroliniana	0.2	0.2	0.1	<0.1	<0.1	0.1
Cornus florida	0.2	0.1	0.3	<0.1	<0.1	<0.1
Taxus canadensis	0.2	0.3		<0.1	<0.1	
Acer negundo	0.1		0.3	<0.1		<0.1
Corylus spp.	0.1	0.1		<0.1	<0.1	
Rock Outcrops	0.1		0.1	<0.1		0.1
Nyssa sylvatica	0.1		0.1	<0.1		<0.1
Vitis riparia	0.1		0.1	<0.1		<0.1
Quercus bicolor	0.1	0.1		<0.1	<0.1	
Picea mariana	0.1	0.1		<0.1	<0.1	
Zanthoxylum americanum	<0.1	<0.1		<0.1	<0.1	
Agricultural Fields	<0.1	<0.1		<0.1	<0.1	

occidentalis, and *Nyssa sylvatica* were reported only from the Till Plains.

Relative Species Weights

Relative frequency data are of limited value because they give the impression that the forests in western New York consisted largely of the widely distributed beech-sugar maple-basswood forest. The rank order of the most widespread (highest relative frequency) three or four species and the species with the greatest RSWs are more or less the same. However, the RSWs approximate what McIntosh (1957) described as importance values, a measure of a species density and biomass. Species which are widespread (have a high relative frequency) don't necessarily have high relative densities and/or large relative biomasses.

The relative species weights (RSWs) (Table 3) temper this impression. Five species have RSWs greater than 5%, which is typical of species importance values in numerous temperate forest communities. In this data set beech, sugar maple, hemlock, basswood, and elms have the highest RSWs.

Although beech and black oak (*Quercus velutina*) had greater frequency on the Till Plains than on the Allegheny Plateau their RSWs were lower on the Plains than the Plateau. As with the frequency data sugar maple, yellow birch, white pine, and hemlock had higher RSWs on the Plateau while basswood, elm, white and black ash had higher values on the Till Plains.

Community Organization

Neither the frequency nor the IV data provide information pertaining to the communities found within the tract since both involve summary data. Two-way indicator species analysis

(TWINSPAN) (Hill 1979a) was used to ascertain the community types. Two analyses were performed; one arranged species, the other samples. Communities were named according to the dominant species (those with RSWs of greater than 10%).

The initial dichotomy in the sample dendrogram (Fig. 2) separated 1,889 samples with beech, sugar maple, and wetland components from 160 samples with oak affinities.

The 1,889 samples were further subdivided into those with wetland components and those from upland mesic sites with beech-sugar maple-hemlock forests. All of the wetland sites contained black ash. Most of the bottomland communities were hemlock-black ash-yellow birch-white pine forests. Eleven samples were in black ash-alder (*Alnus incana*)-elm, eight in northern white cedar (*Thuja occidentalis*)-larch (*Larix laricina*)-alder bottomland forests and three were described by surveyors as black ash swamp containing northern white cedar and white pine.

The most widespread community type was the beech-sugar maple-basswood-elm-hemlock upland forest (1325 samples, 65% of the area). The other beech-sugar maple samples contained components of either oak and white pine or hemlock and yellow birch.

Of the 160 samples (7.8% of the total) with oak affinities, 112 (5.5%) contained mesic species such as sugar maple, basswood, and *Magnolia acuminata* and 48 (2.3%) dominated primarily by oaks. This separation is in part due to the way in which samples were established. A surveyed mile could include mesic communities of beech, sugar maple, or *Magnolia* as well as oak-dominated communities on dry-mesic sites. Although these communities existed on slightly different sites, both were often encountered in the same surveyed mile and, therefore, were combined in the present analysis.

Fig. 3 shows the distribution of communities identified by TWINSPAN analysis. The beech-sugar maple-white ash-elm-hemlock communities (No. 0) were located throughout the tract occurring in almost all of the townships. Those communities with a large component of elm, silver maple, and black ash (No. 2) were concentrated in the northwest quadrant of the tract. A concentration of communities containing black ash, northern white cedar, larch, white pine, and alder, occurred in townships 13 and 14 in ranges I and II (No. 9 in Fig. 3). Today this area includes the Oak Orchard Swamp and the Elba mucklands. A rather large black spruce (*Picea mariana*) swamp existed in townships 4 and 5 in range IX (No. 5 in Fig. 3). This is the Conewango Swamp described in Gordon (1940).

Oak-dominated communities were located in the southeast corner of the tract. A concentration of communities with large components of chestnut and hemlock (Nos. 3, 4, 6, 11 and 13) occurred in the area of the Allegany Reservation. Areas described as "plains" or grassy areas (Range VI and VII, town-

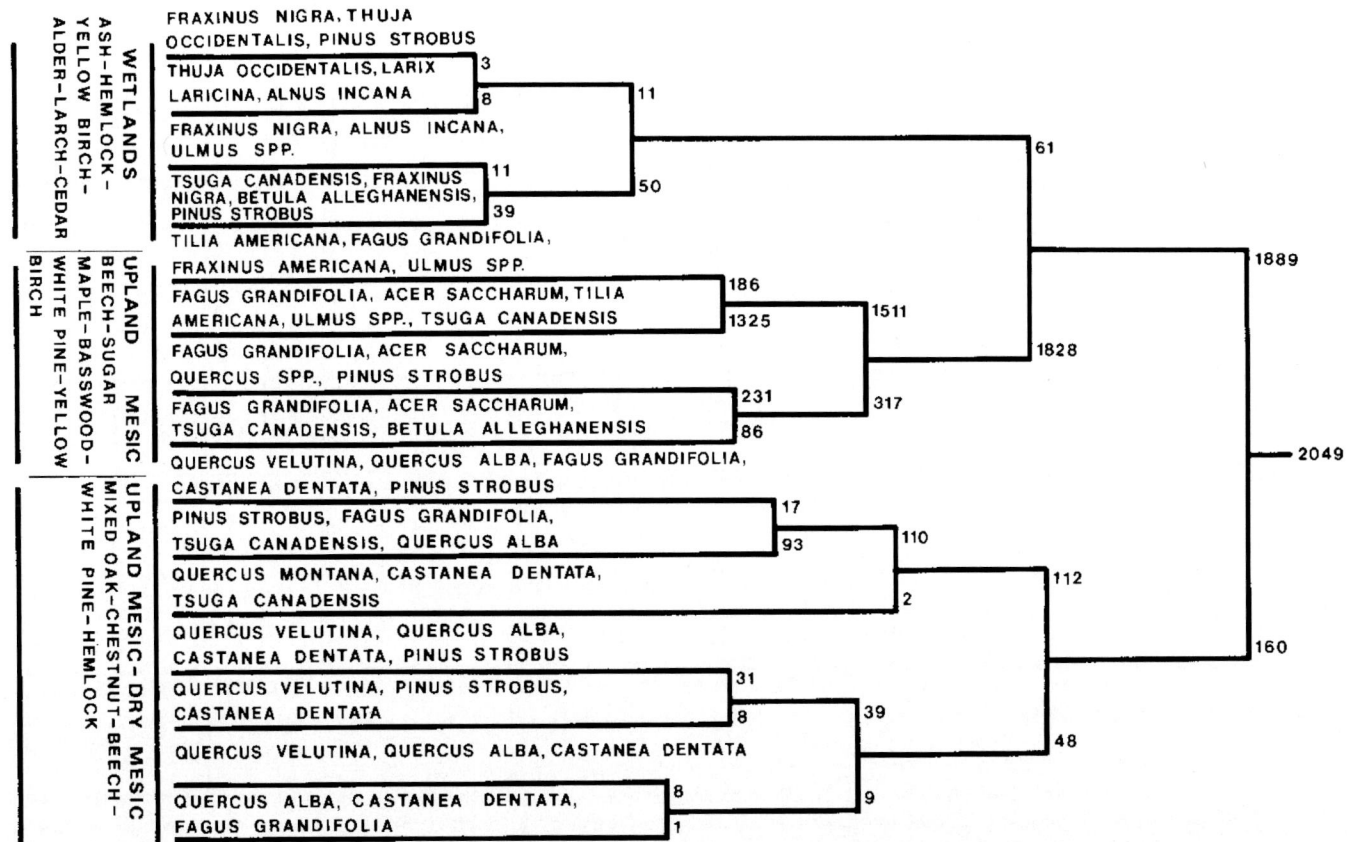

Fig. 2. A dendrogram showing the classification of 2049 surveyed miles. Communities were found in wet and wet-mesic bottomlands, upland mesic, and upland dry-mesic sites. The number of samples included and the dominant species in each community are indicated.

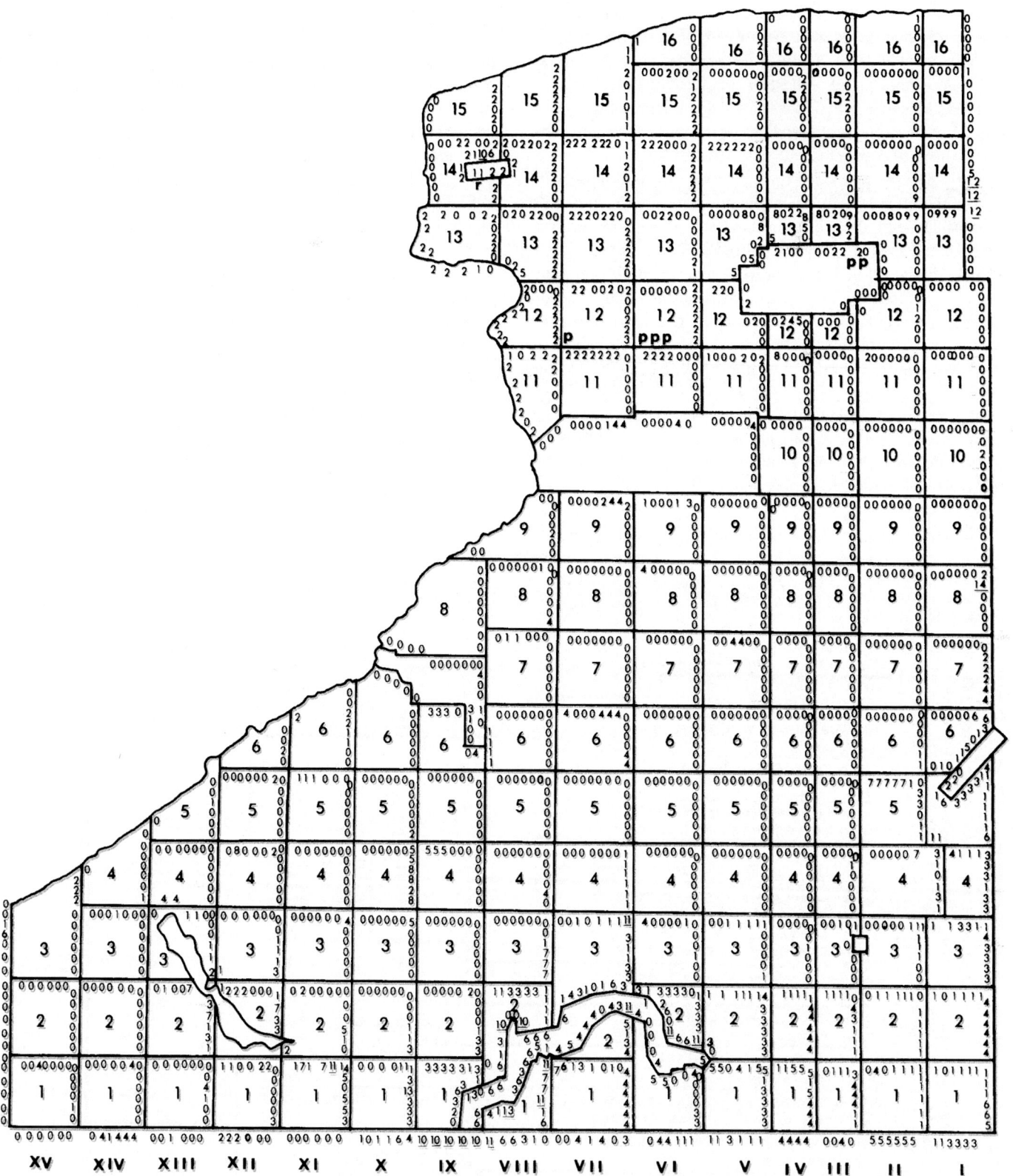

Fig. 3. A map of the community types identified in the sample TWINSPAN analysis (Fig. 2). Large numerals indicate township numbers. Smaller numerals indicate community types shown in Fig. 2 and are designated as follows: 0, beech-sugar maple-white ash-elm-hemlock; 1, beech-sugar maple-oak-white pine; 2, basswood-beech-white ash-elm; 3, white pine-beech-hemlock-white oak; 4, beech-sugar maple-hemlock-yellow birch; 5, hemlock-black ash-yellow birch-white pine; 6, black oak-white oak-chestnut-white pine; 7, black oak-white oak-beech-chestnut-white pine; 8, black ash-alder-elm; 9, northern white cedar-larch-alder; 10, black oak-white pine-chestnut; 11, black oak-white oak-chestnut; 12, black ash-northern white cedar-white pine; 13, chestnut oak-chestnut-hemlock; 14, white oak-chestnut-beech. P designates the location of "plains" and R the location of rock outcrop.

ships 11 and 12); some as being "thinly timbered" and in the literature as "oak openings" (Seischab and Orwig 1991) existed (Fig. 3). Most of these sites were grasslands on droughty soils. Such a site still exists in township 12, Range VII and is on droughty Wassaic soils, 50-70 cm in depth, overlying limestone bedrock. Vegetation surrounding these "plains" were described as a mixture of white, black, and red oak, hickory, "Aspine", ash, ironwood, sugar maple and basswood. Two of these "plains" were described as being "thinly timbered" with the aforementioned species. Another area (Ranges II and III, township 13) described as "plains" was at the edge of the Tonawanda Reservation and was probably an anthropogenic disturbance since that portion of the site which is not presently in agriculture supports vegetation which is clearly mesic (personal observation). A region described as "rock bottom", assumed to be a rock outcrop, was recorded from township 14, range IX.

The species TWINSPAN classification (Fig. 4) resulted in similar community assemblages. The species were grouped into those which occurred on dry-mesic uplands, in wetlands, in wet-mesic bottomlands, or in mesic uplands. These assemblages corroborate the findings based on the sample classification.

Species Distributions

The distribution of several species is clearly associated with the origin of the underlying soils (Fig. 5). Glacial deposits are the main parent materials in this section of western New York (Cline and Marshall 1977). These include glacial till and outwash, deltaic sands, and sediments in former glacial lakes. In addition, on the Genesee-Orleans County line is an accumulation of organic soils, primarily Carlisle Muck. These histosols formed in the Salina Trench, a depression found between the Niagara Escarpment to the north and the Onondaga Escarpment to the south (Fairchild 1928). Lacustrine deposits are concentrated in Niagara and Erie counties in the northwest corner of the area as well as in glacial valleys, in the southern half of the area, which supported lakes as the glacier receded.

Maps of the relative species weights (RSW) were completed (Fig. 6.1—6.12) with circles representing each mile in which a species occurred. The size of the circle indicates the RSW of the species in any particular mile.

Several patterns of species distribution are evident in the maps. The widely distributed beech, sugar maple, and basswood exemplify the first pattern. These occurred on almost all soil types. Beech (Fig. 6.1) reached its greatest RSW at the northeast-

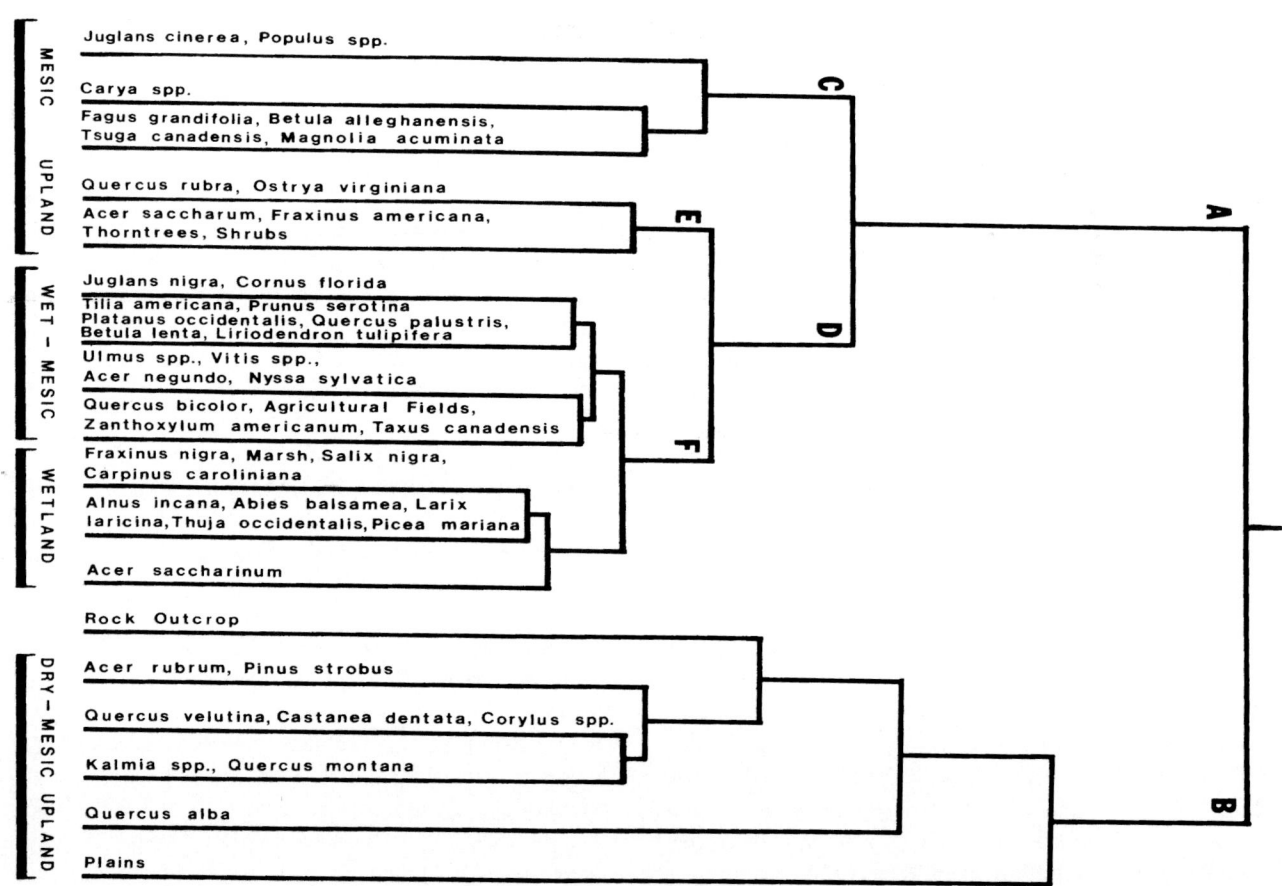

Fig. 4. Species TWINSPAN analysis. Four clusters were identified: species occurring primarily on dry-mesic uplands, wetlands, wet-mesic bottomlands, and mesic uplands. The A includes species generally found in mesic and wet-mesic communities; B includes those on dry and dry-mesic sites. C includes mesic site species associated with beech, yellow birch, and hemlock. D includes mesic site species associated with sugar maple and wet-mesic site species found with silver maple, black ash, and elm. E contains mesic site species and F the bottomland and wetland species.

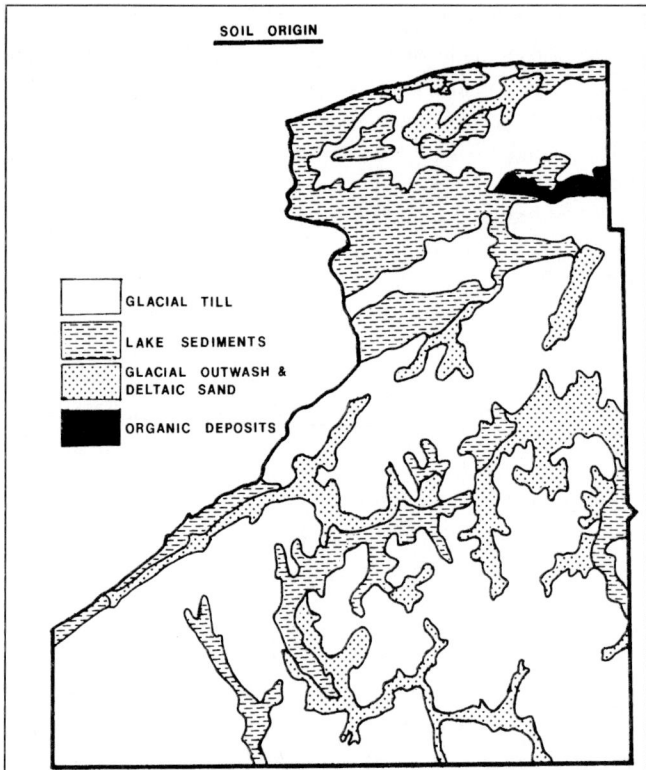

Fig. 5. A soils map indicating the origin of soils within the Holland Company Lands. Areas of glacial till, lake sediment, glacial outwash and deltaic sand, and of organic deposit origin are shown. This map is based on Cline and Marshall (1976).

ern portion of the Allegany Reservation and along Lake Erie. Sugar maple (Fig. 6.2) reached its greatest RSW on the Allegheny Plateau, particularly northeast of the Allegany Reservation and east and southeast of the Buffalo Creek Reservation. Soils in this area were described by Cline and Marshall (1977) as frigid. Basswood (Fig. 6.3), often a component of beech-sugar maple dominated forests was found most often on bottomland sites. Basswood had lower RSWs than either beech or sugar maple in most surveyed miles and was absent from many miles in the southeastern townships. This is where white pine, hemlock, and oaks dominated.

A second distributional pattern is characterized by the oaks. Black and red oak (Fig. 6.4) were distributed in the southern townships, particularly in the southeast and along the edges of the Allegany Reservation. Soils in this region were described as having cambic horizons with low base status (Dystrochrepts) (Cline and Marshall 1977). Most of these soils were described as fine loamy soils of sandstone and siltstone frost-churned residuum. These two species were also on the Lake Erie and the broader Lake Ontario Till Plain. They were notably absent from the center of the tract, northeast of the Allegany Reservation and east and southeast of the Buffalo Creek Reservation where sugar maple had its greatest RSWs.

White oak (Fig. 6.5) had a distribution similar to black and red oak, occurring in the southeastern townships and on the Till Plain, particularly the Lake Ontario Plain. It was notably concentrated along the Niagara River from the Buffalo Creek Reservation to Lake Ontario. A similar concentration of white ash (Fig. 6.9) lay along the river. A white oak-white ash community still exists along the bluff overlooking the Niagara River and is visible along the Robert Moses State Parkway (personal observation). White oak was missing from the record for the central portion of the tract where sugar maple and beech dominated. Oaks weren't found to be significantly associated with soils containing a fragipan as shown by Whitney (1990) on the Allegheny Plateau nearby in Pennsylvania.

American Chestnut (*Castanea dentata*) (Fig. 6.6) was a significant component of oak-dominated forests, particularly in the southeastern townships. It reached its greatest RSW around the Allegany Reservation.

Black ash (Fig. 6.7) occurred principally in wetlands and bottomland forests and represents another pattern of distribution. The greatest concentration of communities described as black ash swamp or black ash-elm-maple (presumably silver maple) was on the Lake Ontario Till Plain. A large concentration occurred in the vicinity of the Tonawanda Reservation. Distribution of this species was closely associated with that of soils developed from lake sediments (Fig. 5) as well as the histosols on the Genesee-Orleans County line. In the southern half of the tract, a concentration of black ash occurred in township 4 of ranges IX and X, an area known as Conewango Swamp (Gordon 1940), again on soils of lacustrine origin. Since many surveyor line descriptions included black ash-elm, one might expect the distribution of elm (Fig. 6.8) to be similar to black ash; however, it's quite different. Surveyors did not distinguish between *Ulmus americana*, a bottomland species, and *U. rubra*, an upland species. Consequently *Ulmus* has a wide distribution throughout the tract, although it was clearly less abundant in the southeastern townships.

White ash was a component of beech-maple, oak, and bottomland forests (Fig. 6.9). It was absent, however, northeast of the Allegany Reservation, a pattern similar to the oaks. It was rather sparse in the southeast townships where hemlock and white pine dominated. Its greatest concentration was on the bluff overlooking the Niagara River.

The two conifers most often mentioned were hemlock (Fig. 6.10) and white pine (Fig. 6.11). Hemlock was a component of some wetlands, ravine communities, as well as upland forests (Fig. 6.10). Its largest RSWs occur along Lake Erie south of the Cattaraugus Reservation on lacustrine soils and in the vicinity of the Caneadea Reservation on the Genesee River. It was notably absent on the soils of lacustrine origin in the northwest corner of the area. White pine (Fig. 6.11) occurred in southern townships and in wetlands or bottomlands in other locations. Its greatest RSWs were in the vicinity of the Allegany Reservation where it was part of the black oak-white oak-chestnut-white pine community. At the edge of the Caneadea Reservation it was the leading dominant in the white pine-beech-hemlock-white oak community. This same community is found today in the ravines tributary to the Genesee River, particularly in Letchworth State Park. The southern townships were in a dissected landscape with a dendritic pattern of streams. Many of the slopes, particularly

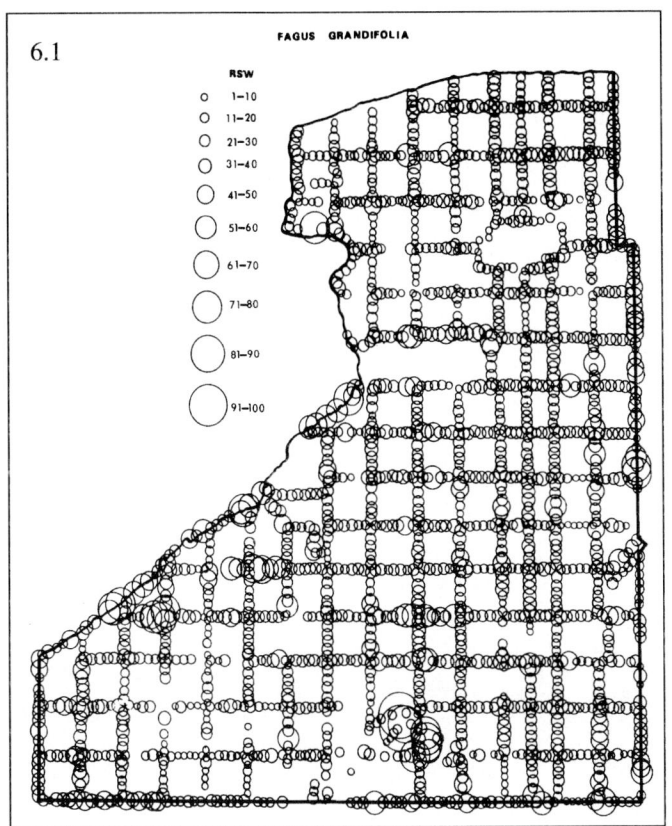
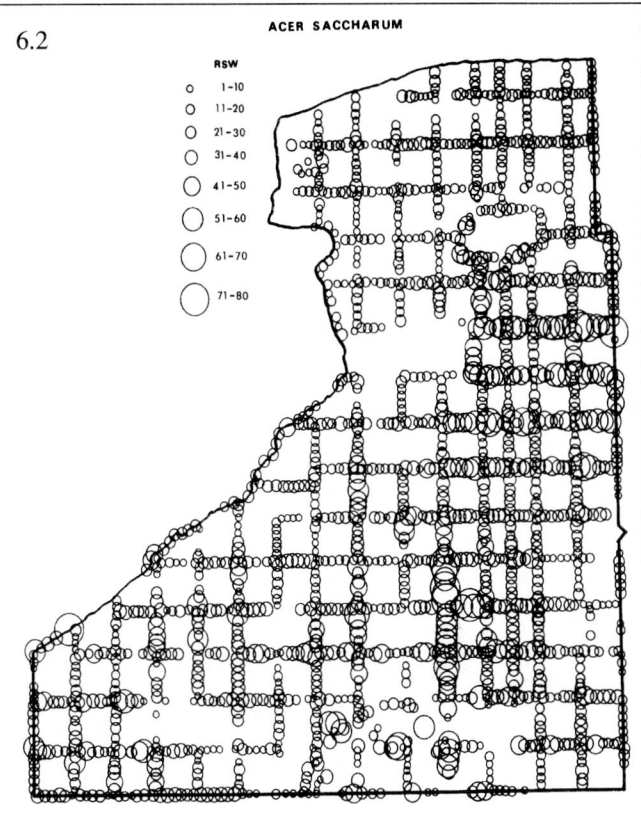
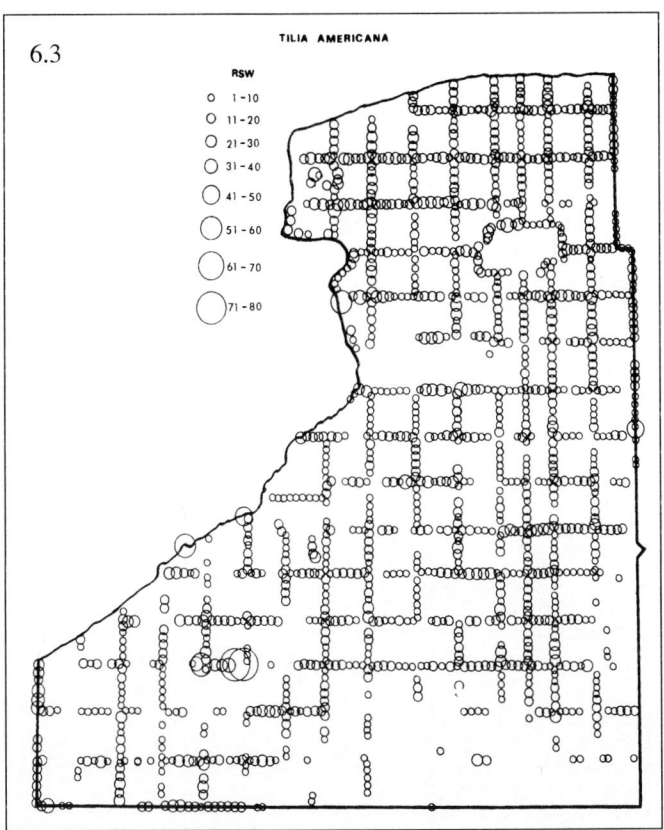
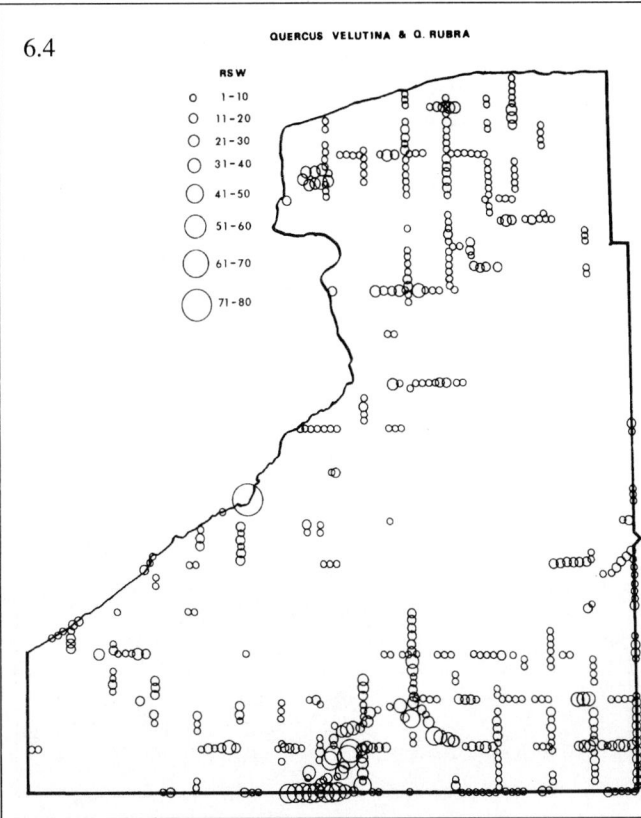

Figs. 6.1-6.12. Species distribution maps along township lines. Circles represent the relative species weight (RSW) of the species in each surveyed mile in which it occurred.

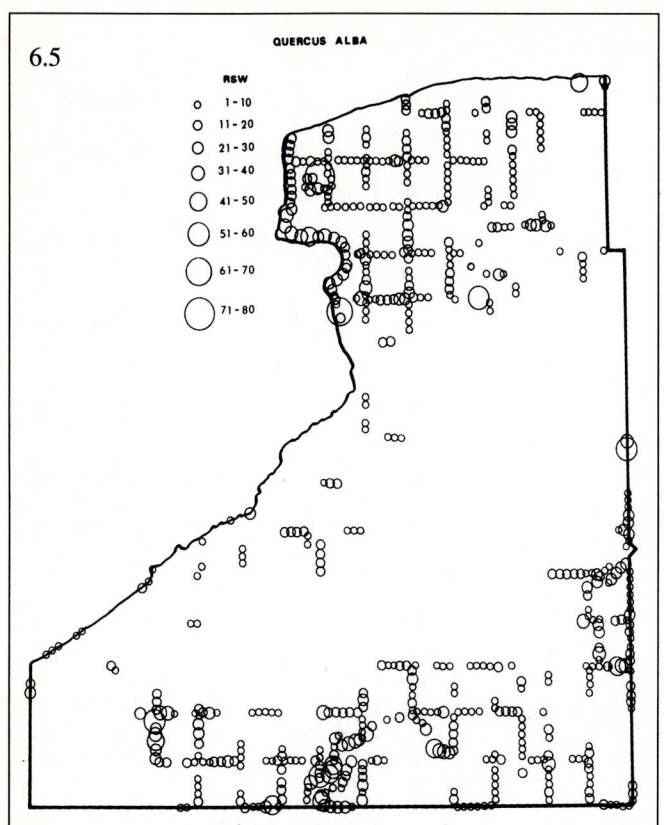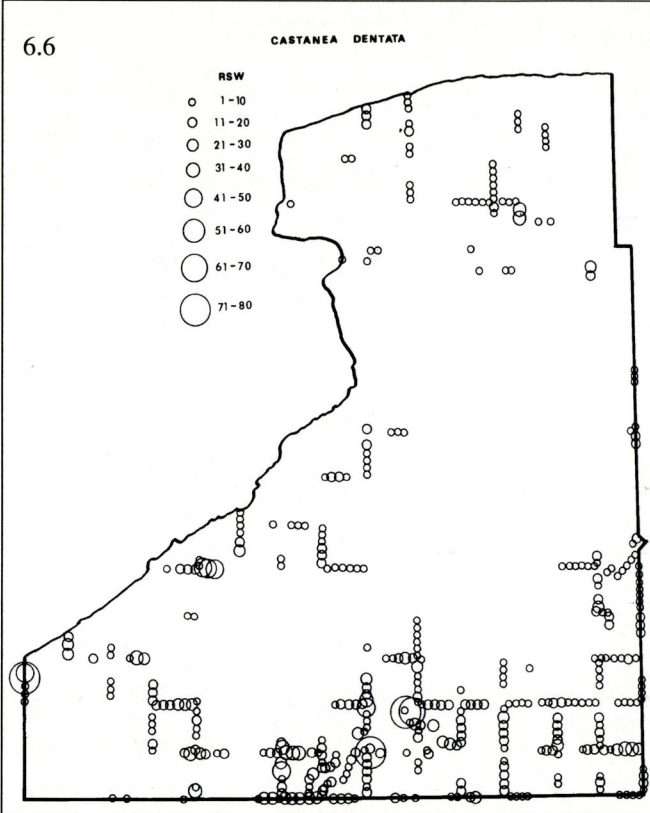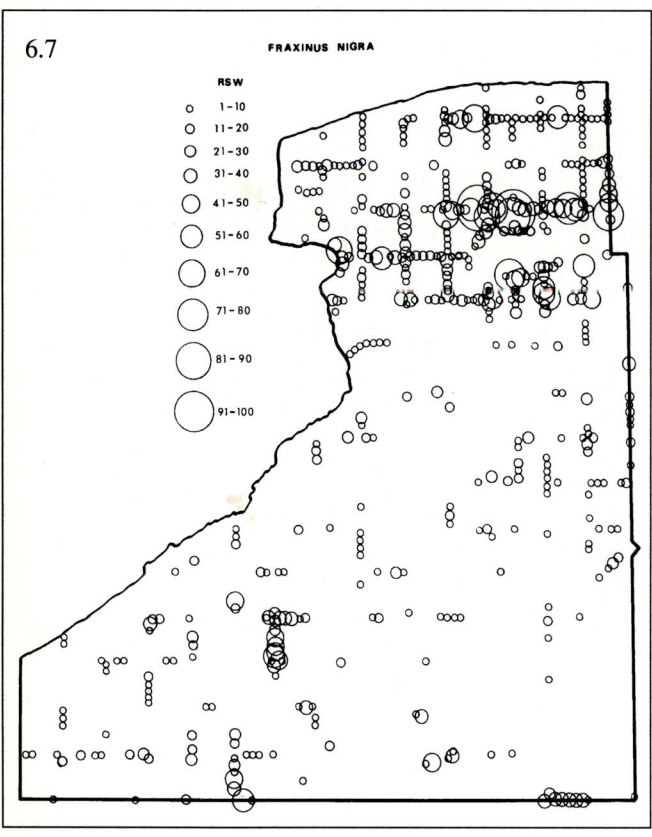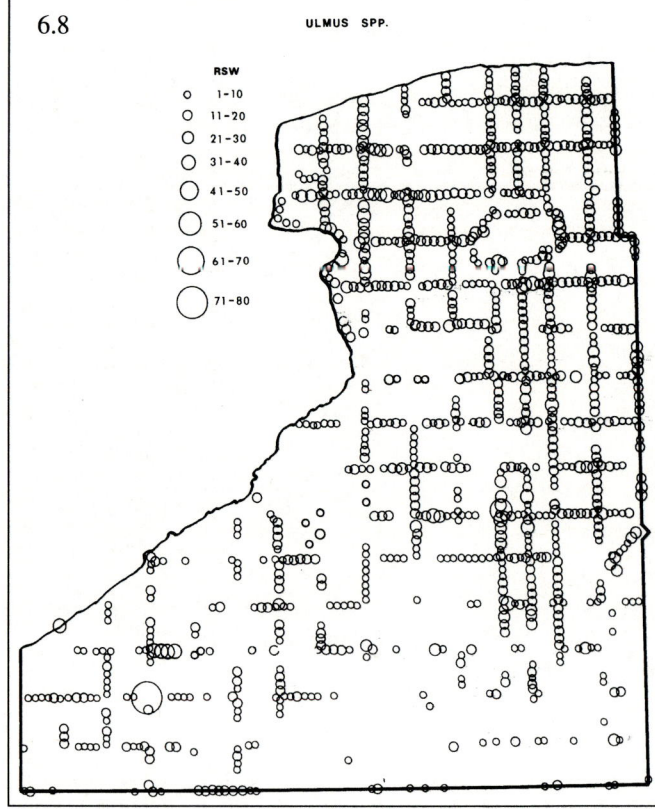

Fig. 6. (continued)

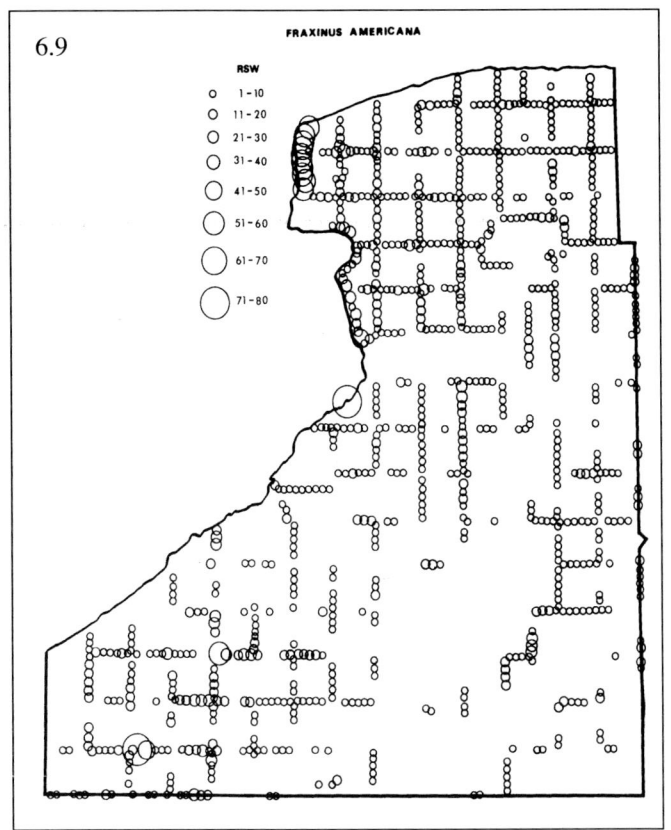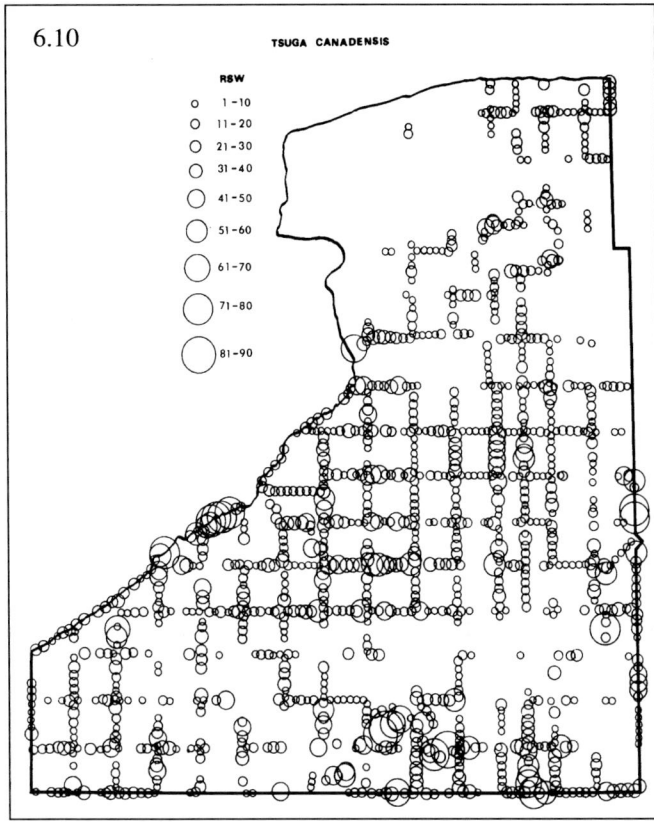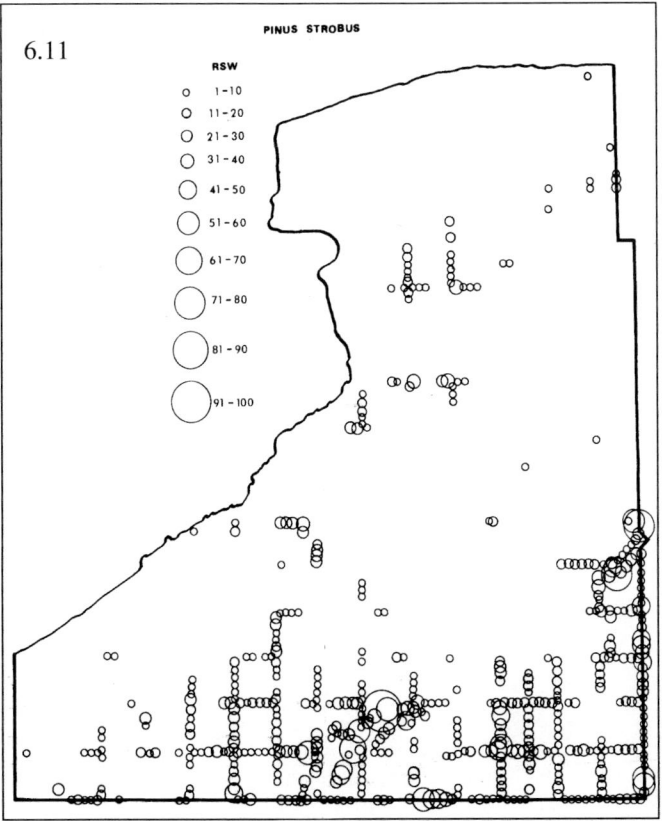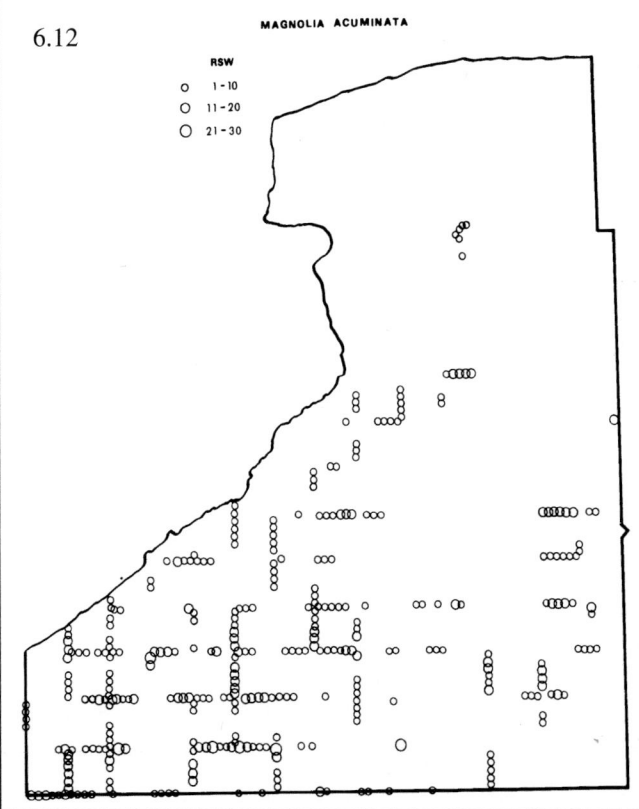

Fig. 6. (concluded).

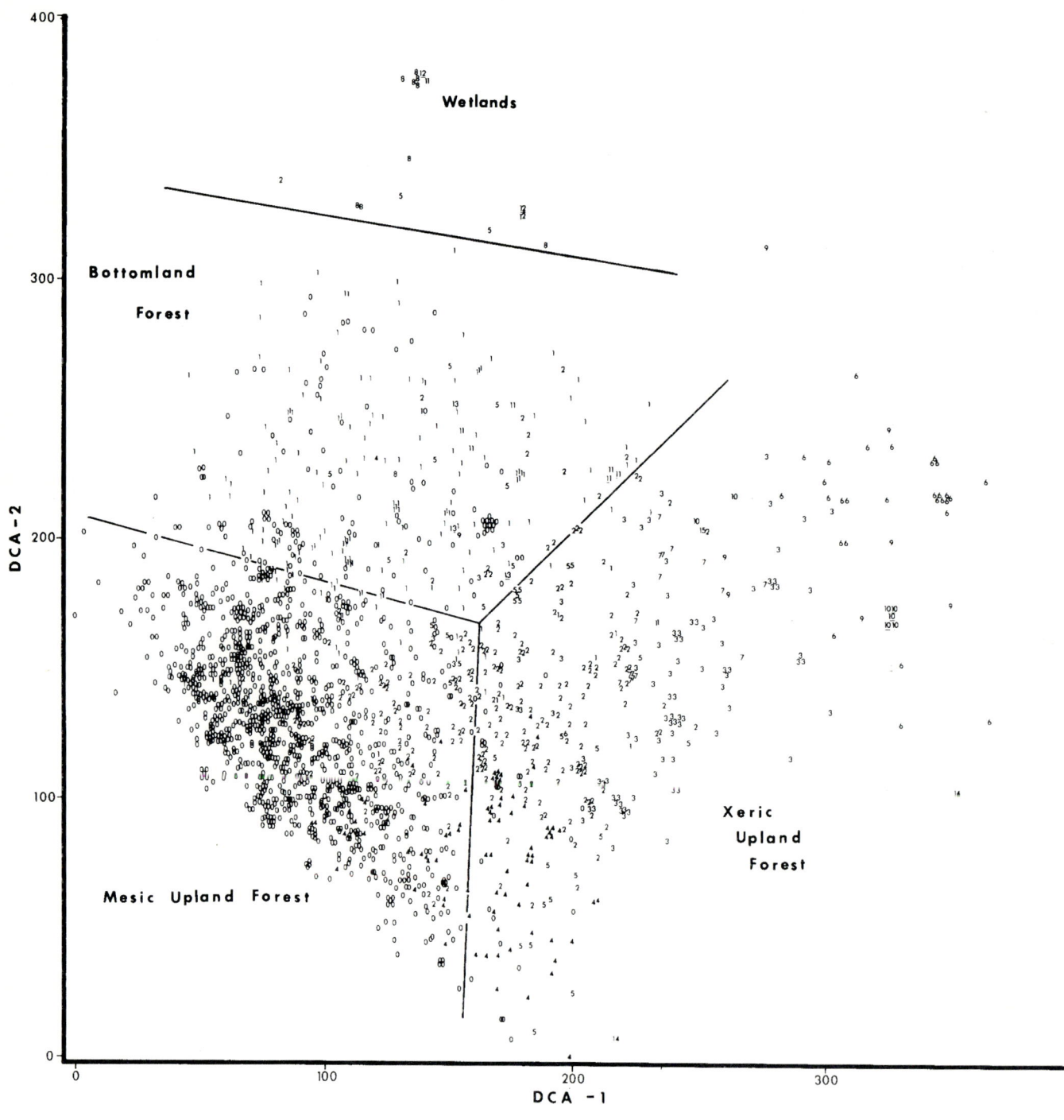

Fig. 7. Sample detrended correspondence analysis ordination with vegetation groups indicated. Numbers represent communities dominated by: 0, beech-sugar maple-basswood-butternut-cucumber tree; 1, black ash-silver maple-elm-hemlock; 2, black oak-red oak-chestnut-hemlock-white pine; 3, chestnut-red maple-cucumber tree-poplar; 4, beech-white oak-butternut; 5, hemlock-yellow birch-sugar maple; 6, chestnut-sugar maple-hickory-chestnut oak; 7, hemlock-black cherry; 8, alder-yellow birch; 9, beech-red maple-black oak; 10, white oak-sugar maple; 11, beech-alder-larch-white ash; 12, black ash; 13, black spruce; 14, chestnut oak; 15, red maple.

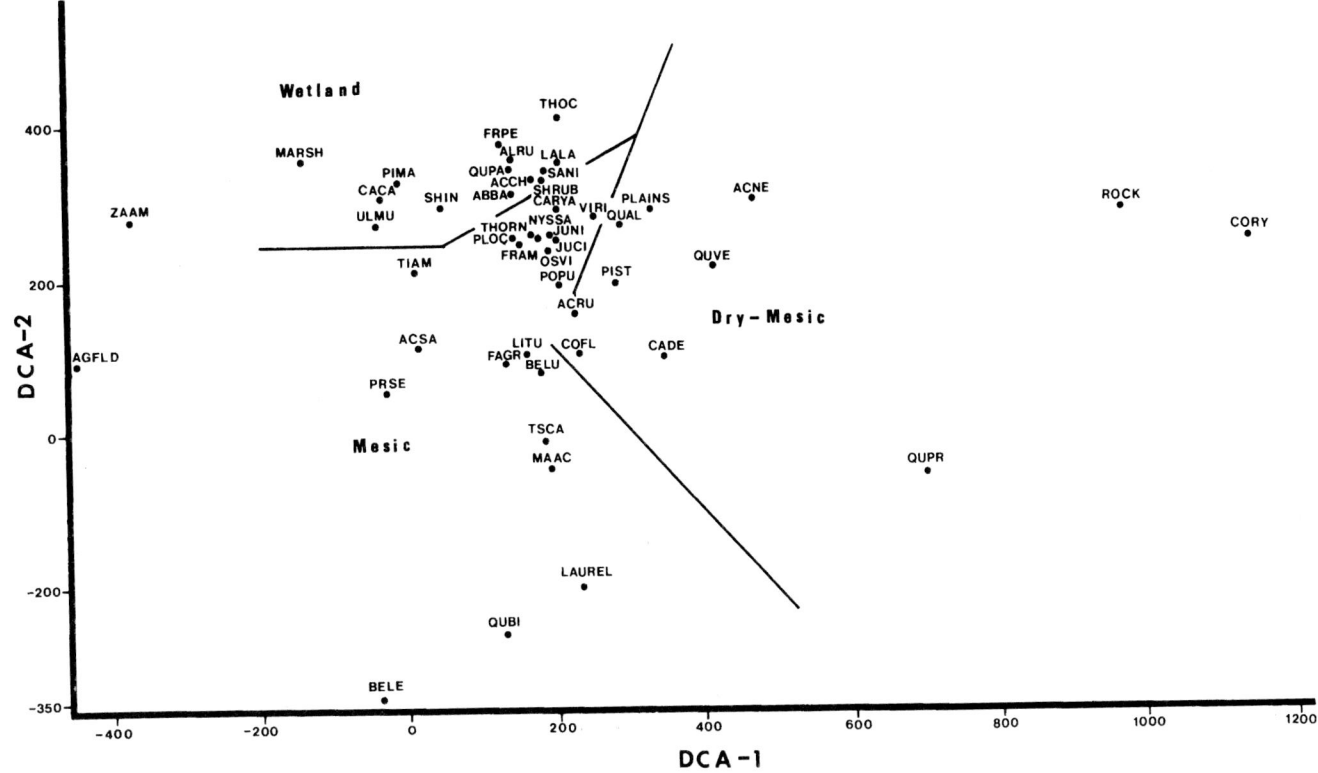

Fig. 8. Species DCA ordination. Species designations are as follows: ABBA, *Abies balsamea*; ACRU, *Acer rubrum*; ACSA, *A. saccharum*; ACCH, *A. saccharinum*; ACNE, *Acer negundo*; AGFLD, Agricultural Fields; ALRU, *Alnus incana*; BELE, *Betula lenta*; BELU, *B. alleghanensis*; CADE, *Castanea dentata*; CACA, *Carpinus caroliniana*; CARYA, *Carya* spp.; COFL, *Cornus florida*; CORY, *Corylus* spp.; FAGR, *Fagus grandifolia*; FRAM, *Fraxinus americana*; FRPE, *F. nigra*; JUCI, *Juglans cinerea*; JUNI, *J. nigra*; LALA, *Larix laricina*; LITU, *Liriodendron tulipifera*; MAAC, *Magnolia acuminata*; MARSH, marshes and mires; NYSSA, *Nyssa sylvatica*; OSVI, *Ostrya virginiana*; PIMA, *Picea mariana*; PIST, *Pinus strobus*; PLAINS, grassland; PLOC, *Platanus occidentalis*; POPU, *Populus* spp.; PRSE, *Prunus serotina*; QUAL, *Quercus alba*; QUPA, *Q. palustris*; QUBI, *Q. bicolor*; QUPR, *Q. montana*; QUVE, *Q. velutina*; LAUREL, *Kalmia* spp.; ROCK, rock outcrop; SANI, *Salix nigra*; SHIN, *Taxus canadensis*; THORN, *Crataegus* spp.; THOC, *Thuja occidentalis*; TSCA, *Tsuga canadensis*; ULMUS, *Ulmus* spp.; VIRI, *Vitis riparia*; ZAAM, *Zanthoxylum americanum*.

north facing, coves, and ravines contain white pine today (personal observation). A similar pattern of distribution occurred in these forests in the 1790s.

Cucumber tree (*Magnolia acuminata*) (Fig. 6.12) occurred primarily in the southwestern quadrant of the tract. Although this species has been associated with the mixed mesophytic forest that others (Gordon 1940, Braun 1950) described on the unglaciated section of the Allegheny Plateau (south of the Allegheny River), the species was noticeably sparse in the survey notes from this area.

Phytosociology and Environmental Gradients

The DCA ordination of samples separates wetland, bottomland, mesic upland, and xeric upland forests (Fig. 7). Wetland communities were dominated by alder, black ash or larch and are at the top of the ordination. In a lower position on the second axis are the bottomland forests of black ash-elm-silver maple. In the lower left quadrant are the mesic upland beech-sugar maple forests. In the lower right quadrant are the xeric upland forests dominated by oaks, chestnut and other species, mostly from upper slopes.

The ordination indicates a vegetational response to two interdependent environmental gradients. The first axis correlates with a topographic gradient, and the second axis correlates with a moisture gradient. Those sites at the lower left were on the lower slopes and those in the lower right were on the middle to upper slopes. Some, at the right, were also on the higher and drier flat areas above the Niagara River and at the Tuscarora Reservation.

The species DCA ordination (Fig. 8) arranged the species in a similar manner. The second axis implies a moisture gradient with wet site species (cedar, black ash, alder, larch and marshes) at the top and drier site species (white oak, chestnut, chestnut oak) further down and to the right on the first axis. Mesic species (beech, sugar maple, yellow and black birch, hemlock) are at the center and lower portion of the ordination.

Disturbance

These forests were not all in a climax state. Catastrophic disturbances had occurred and were recorded by surveyors (Seischab and Orwig 1991). Windthrow had caused the greatest amount of disturbance, 10.4 kilometers of surveyed lines having been noted as "windthrow" or "downed timber". These were all recorded from the Allegheny Plateau and comprised 0.5% of the area. Such disturbance was probably due to thunderstorms, possibly to tornadoes. A contributing factor was also trees thrown during glaze storms. Most occurrences of windthrow noted in

Table 4. Relative frequency (RF) and Relative Species Weights (RSW) on the Allegheny Plateau and Till Plain of the Holland Company Lands and the Phelps and Gorham Purchase.

	HOLLAND LAND COMPANY				PHELPS AND GORHAM PURCHASE			
	PLATEAU		TILL PLAINS		PLATEAU		TILL PLAINS	
	(RF)	(RSW)	(RF)	(RSW)	(RF)	(RSW)	(RF)	(RSW)
Fagus grandifolia	92.1	23.4	93.1	20.5	56.8	19.0	78.2	32.1
Acer saccharum	84.5	22.6	81.9	15.6	53.3	12.3	71.1	18.4
Tilia americana	61.1	7.1	84.7	10.5	25.7	3.0	59.7	12.2
Tsuga canadensis	64.5	11.4	41.9	7.3	36.1	11.1	16.9	5.0
Ulmus spp.	43.4	4.7	71.6	9.1	16.4	2.4	51.6	7.7
Fraxinus americana	33.4	2.6	69.7	6.3	21.5	3.1	49.8	8.4
Betula alleghanensis	39.2	3.7	11.7	0.8	11.0	1.3	5.1	0.5
Quercus alba	25.7	3.6	39.7	4.2	42.7	10.3	42.8	10.6
Fraxinus nigra	13.3	1.7	45.3	7.3	4.4	0.6	28.0	5.2
Pinus strobus	34.2	4.7	8.7	0.9	38.3	10.8	7.9	2.2
Quercus velutina	18.3	2.6	25.9	1.9	35.8	9.3	38.5	9.0
Castanea dentata	25.0	3.1	11.4	0.9	27.0	5.6	10.6	2.4
Carya spp.	7.6	0.6	43.0	2.9	15.6	2.6	27.8	5.2
Acer rubrum	19.4	2.1	21.4	2.5	6.3	0.7	3.7	0.6
Prunus serotina	16.5	1.1	0.3	6.4	4.3	0.3	3.2	0.2

the survey had occurred on steep slopes. Recent evidence has shown a correlation between steepness of slope and the percent of trees toppled during glaze storms (Seischab *et al*, in review).

Catastrophic windthrows are quite common in forests of the northeast. Canham and Loucks (1984) reported return times of 1000 years for windthrow in Wisconsin. Bormann and Likens (1979) indicated that large scale disturbances in the forests of the White Mountains were principally due to windthrow with "very little evidence, vegetationally or historically, that fire was widespread". Windthrow evidence has been recorded for both bottomland (Whitney 1986) and upland forests in both conifer (Lorimer 1977) and deciduous communities.

As in the forests of the White Mountains, the Holland Company survey records provided no evidence of fire as a disturbance. In the Phelps and Gorham Purchase to the east there was one record of fire in the original survey (Seischab and Orwig 1991). That record was at the edge of a pitch pine (*Pinus rigida*) stand where one might expect to find fire evidence. The second survey of that tract indicated numerous fires in the Town of Wheatland. Since the first survey reported mature forest along

Table 5. a. A comparison of forest data from New York and adjacent Pennsylvania. Data included are those from 1. the Catskill Mountains (McIntosh 1962); 2. the Military Tract (Marks and Gardescu, present volume); 3. the Phelps and Gorham Purchase (Seischab 1990); 4. the Allegheny Plateau of Pennsylvania (Whitney 1990), and 5. Pennsylvania (Lutz 1930). Both Allegheny Plateau and Till Plain data are shown. Both relative frequency (RF) data and percent of witness tree (% of trees) data are shown. b. Coefficients of similarity between the above mentioned tracts.

a.

	Catskill (% of trees) (1)	Military (RF) (2)	P&G Plat. (RF) (3)	P&G Plain (RF) (3)	HLC Plat. (RF)	HLC Plain (RF)	Penn. Plat. (% of trees) (4)	Penn. (% of trees) (5)
Beech	49.5	72.0	56.8	78.2	92.1	93.1	43.4	30.9
Hemlock	20.3	19.2	36.1	16.9	64.5	41.9	19.9	26.8
Sugar Maple	12.8	15.7	53.3	71.1	84.5	81.9	5.3	8.1
Maple & Red Maple		52.6	6.3	3.7	19.4	21.4	4.7	5.0
Basswood	1.3	47.3	25.7	59.7	61.1	84.7	0.4	0.1
Birch spp.	7.3	3.4	12.4	5.8	39.6	12.3	6.3	6.1
White Pine	0.5	9.6	38.3	7.9	34.2	8.7	3.1	6.0
Chestnut	0.5	5.5	27.0	10.6	25.0	11.4	2.8	5.6
White Oak		7.9	42.7	42.8	25.7	39.7	4.1	0.6
Red, Black & Scarlet Oak	0.3	6.5	38.0	38.7	23.5	30.0	0.6	0.2
Other Oaks		11.3	2.5	0.5	1.6	2.3	0.4	2.7

b.

	Coefficients of Similarity				
	Catskill (1)	Military	P & G	HLC	Penn. Plat.
Military Purchase (2)	42.9				
Phelps & Gorham Purchase (3)	41.1	65.0			
Holland Land Company	46.5	66.7	84.4		
Allegheny Plateau, PA (4)	79.9	51.3	44.2	50.0	
Pennsylvania (5)	51.1	47.5	55.5	60.5	54.8

the same survey line, it can be concluded that fire was used in the clearing of the forests for agricultural production.

Comparison of Forests, Circa 1749-1815

The Till Plain and Allegheny Plateau forests of the Phelps and Gorham Purchase (Seischab 1990), beginning 19 km to the east, differ somewhat from the Holland Land Company tract (Table 4). Although beech, sugar maple, basswood, and hemlock were widely distributed on the Allegheny Plateau of both tracts they were more abundant in the western tract. The Plateau of the Phelps and Gorham Purchase had greater frequencies and larger RSWs of white oak, black oak, and white pine with a larger oak-pine component. The Till Plains of the Holland tract had higher frequencies of beech, sugar maple, basswood, hemlock, elm, ash, and yellow birch. The Till Plain of the Phelps and Gorham had more white and black oak, which occurred on sandy outwash areas in Monroe County, in the northwest portion of the Phelps and Gorham Purchase.

The forests on the Till Plain of the Holland Land Company contained less hemlock and more basswood than did the Allegheny Plateau. The Allegheny Plateau portion of the Holland Land Company is similar to those originally surveyed in the Catskill Mountains to the east (McIntosh 1962) which was also dominated by beech, hemlock and sugar maple (Table 5a), however, basswood was not a dominant species in the Catskills. South of the Holland Company Lands, on the Allegheny Plateau in Pennsylvania, the forests were again dominated by beech, hemlock, and sugar maple (Lutz 1930, Whitney 1990), again lacking in significant amounts of basswood. These Pennsylvania forests and those of the Catskills were very similar, being dominated by beech, hemlock, sugar maple, and birch, and containing very little basswood. Basswood was more characteristic in central and western New York forests where it had a relative frequency between 25.7% and 84.7% (Table 5a).

In the Military Tract (Marks and Gardescu, present volume) the top-ranking bounds taxa on the Plateau were beech, maple, and basswood followed by oaks (including white), whereas on the Lowland they were beech, maple, basswood, then hemlock. Thus the Military Tract seems to have been more similar to the forests of western New York than those of adjacent Pennsylvania.

The similarity coefficients indicate that the vegetation of the Holland Company was most similar to that of the Phelps and Gorham purchase (84.4%) immediately to the east (Table 5b). It was 66.7% similar to the vegetation of the Military Purchase and only 50-60% similar to the forests in Pennsylvania to the immediate south. The forest of the Allegheny Plateau in Pennsylvania (Whitney 1990) were most similar to those of the Catskill Mountains (McIntosh 1962) (79.9%) rather than to those of the Holland Company Lands to their immediate north.

DISCUSSION

Gordon (1940) examined the primeval forests of Cattaraugus County (one of the eight counties included in the present study), on the Pennsylvania border of the Holland Company Lands, using bearing trees identified in the survey notes to identify "edaphic climax associations" as described by Weaver and Clements (1929) and later used by Braun (1950). He identified six associations: Oak-Chestnut Forest on dry ridges, south and southwest facing slopes, Mixed Mesophytic Forest on middle to upper slopes, the Beech-Sugar Maple Forest lacking hemlock and birch on the better drained soils near ridge tops, Bottomland Hardwood Forests along the major tributaries, White Pine-American Elm Swamp Forest on "river flats" and floodplains, and Black Spruce-Tamarack Bog Forest on organic soils in depressions of glacial origin.

In a general way, these association names can be used in the classification of forests in the rest of the Holland Company Lands, recognizing that community demarcations were usually not clearly defined.

The occurrence of beech-maple, oak-chestnut, oak-hickory, and bottomland forests in the Holland Company Lands are, generally, as described by Gordon (1940). Those forests which he described as White Pine-American Elm Forest were more often described by surveyors as a combination of black ash, elm, and silver maple forests in most of these western counties. White pine was more often a component of bottomland or wetland forests which included hemlock and, at times, northern white cedar. These conifer swamps occurred less frequently than did the black ash-elm-silver maple forests.

Braun (1950) recognized two forest types in western New York: Beech-Maple forests on the Till Plain north and west of the Allegheny Plateau, and Hemlock-White Pine-Northern Hardwoods forest on the Allegheny Plateau. The beech-maple communities in this study lay primarily on the Lake Plains in agreement with Braun (1950). They also included ash-silver maple-elm swamp forests, particularly in Niagara and Orleans Counties adjacent to Lake Ontario. The beech-sugar maple forests on the Allegheny Plateau were part of the Hemlock-White Pine-Northern Hardwoods Region with hemlock occurring in ravines such as those described by Lewin (1974) for the Finger Lakes area. Hemlock also occurred in the southern tier of counties in the Holland Company Lands. Allegany, Cattaraugus, and Chautauqua Counties, adjacent to Pennsylvania, have extensive areas with dendritic drainage patterns, steep slopes, and steep stream channels. Hemlock and white pine were widely distributed in this area, occurring in coves, ravines, and wetland forests, as well as being a component of the surrounding upland forests.

Gordon (1940) described mixed-mesophytic forest from the unglaciated section of western New York, south of the Allegheny River as occurring between hemlock-beech or beech-sugar maple on lower slopes and oak-chestnut on upper slopes. Similarly, Braun (1950) described such forests occurring at "an intermediate position on slopes between the beech-maple below and the oak-chestnut above." Mixed-mesophytic forest implies a great diversity of oaks and such characteristic species as *Magnolia acuminata*, *Nyssa sylvatica*, and *Liriodendron tulipifera*, as well as the dominants *Fagus grandifolia*, *Acer saccharum*, *Tsuga canadensis*, and *Acer rubrum*.

Data from the forests of the 1790s do not demonstrate a great diversity of oaks in the unglaciated section of western New York, nor an extensive presence of *Magnolia*, *Nyssa*, or *Liriodendron*. The average number of species recorded/mile surveyed for the Holland Company Lands was 6.75. That for the unglaciated section was 5.42 species/mile, indicating a somewhat lower species diversity for the unglaciated section. The survey data do support Gordon's contention of the widespread existence of this community south of the Allegheny River. In a separate TWINSPAN classification of the unglaciated region, beech-sugar maple, hemlock-birch, white pine-red maple, basswood-magnolia-butternut, black oak-chestnut, white ash-white oak-hickory, and wetland forests of black ash-elm or black spruce communities were identified. Even though *Magnolia* was identified as a component of one of these communities, it was more widely distributed to the west of this region than on the unglaciated section. The survey notes do not provide evidence of mixed mesophytic forest in the area.

ACKNOWLEDGMENTS

Many individuals and organizations are recognized for their assistance and support in the generation of this document and those preceding it. Franciska Safran was of tremendous assistance in the investigation of the records of the Holland Land Company. I thank E.H. Ketchledge and S. Gardescu for their assistance in the identification of species names used during the survey period. Comments by J.M. Bernard, P.L. Marks, and T.A. Siccama on an earlier version of this paper and the two reviewers of the present paper greatly improved the manuscript and are much appreciated. I thank Dr. John Paliouras for my having been selected a Dean's Summer Fellow, and the Rochester Institute of Technology for my being granted a sabbatical leave to complete the data gathering for this paper. The Rochester Museum and Science Center is recognized for its grants in support of my projects.

LITERATURE CITED

Bazzaz, F.A. 1975. Plant species diversity in old field successional ecosystems in southern Illinois. Ecology 56: 485-488.

Bormann, F.H. and G.E. Likens. 1979. Catastrophic disturbance and the steady state in northern hardwood forests. American Scientist 67: 660-669.

Braun, E.L. 1950. Deciduous forests of eastern North America. McGraw-Hill Book Co., New York. 596 pp.

Britton, N.L. and A. Brown. 1913. An illustrated flora of the northern United States, Canada and the British Possessions, 2nd ed. Scribner's Sons, New York. (Republished in 1970 by Dover, New York.)

Canham, C.D. and O.L. Loucks. 1984. Catastrophic windthrow in presettlement forests of Wisconsin. Ecology 65: 803-809.

Cline M.G. and R.L. Marshall. 1976. General soil map of New York State. United States Department of Agriculture, Soil Conservation Service.

Cline, M.G. and R.L. Marshall. 1977. Soils of New York Landscapes. Cornell Cooperative Extension Publication, Information Bulletin 119. p. 62.

Fairchild, H.L. 1928. Geologic story of the Genesee Valley of western New York. Rochester, New York. p. 125.

Gordon, R.B. 1940. The primeval forest types of southwestern New York. New York State Museum Bulletin 321: 3-102.

Hill, M.O. 1979a. TWINSPAN. A FORTRAN program for arranging multivariate data in an ordered two-way table by classification of the individuals and attributes. Cornell University, Ithaca, New York. 90 pp.

Hill, M.O. 1979b. DECORANA. A FORTRAN program for detrended correspondence analysis and reciprocal averaging. Cornell University, Ithaca, New York. 52 pp.

Lewin, D.C. 1974. The vegetation of the ravines of the southern Finger Lakes, New York region. American Midland Naturalist 91: 315-342.

Loeb, R.E. 1987. Pre-European settlement forest composition in east New Jersey and southeastern New York. American Midland Naturalist 118: 414-423.

Lorimer, C.G. 1977. The presettlement forest and natural disturbance cycle of northeastern Maine. Ecology 58: 139-148.

Lutz, H. J. 1930. Original forest composition in northwestern Pennsylvania as indicated by early land survey notes. Journal of Forestry 28: 1098-1103.

McIntosh, R.P. 1957. The York Woods, a case history of forest succession in southern Wisconsin. Ecology 38: 29-37.

McIntosh, R.P. 1962. The forest cover of the Catskill Mountain region, New York, as indicated by land survey records. American Midland Naturalist 68: 409-423.

Mitchell, R.S. 1986. A checklist of New York State plants. New York State Museum Bulletin 458. 272 pp.

Pieterse, W.C. 1976. Inventory of the archives of the Holland Land Company. Municipal Printing Office of Amsterdam. 75 pp.

Safran, F.K. 1988. The preservation of the Holland Land Company records. New York History 69: 163-183.

Seischab, F.K. 1990. Presettlement forests of the Phelps and Gorham Purchase in Western New York. Bulletin of the Torrey Botanical Club 117: 27-38.

Seischab, F.K. and D. Orwig. 1991. Catastrophic disturbances in the presettlement forests of western New York. Bulletin of the Torrey Botanical Club 118: 117-122.

Seischab, F.K., J.M. Bernard, and M.D. Eberle. Glaze storm damage to western New York forest communities. *In review*.

Siccama, T.A. 1971. Presettlement and present forest vegetation in northern Vermont with special reference to Chittenden County. American Midland Naturalist 85: 153-172.

Sorenson, T. 1948. A method of establishing groups of equal amplitude in plant sociology based on similarity of species content. Kong. Dan Vidensk. Selsk. Biol. Skr. 5: 1-34.

Weaver, J.E. and F.E. Clements. 1929. Plant Ecology. McGraw-Hill Book Company, New York.

White, C.A. 1984. A history of the rectangular survey system. United States Department of Interior, Bureau of Land Management.

Whitney, G.G. 1986. Relation of Michigan's presettlement pine forests to substrate and disturbance history. Ecology 67: 1548-1559.

Whitney, G.G. 1990. The history and status of the hemlock-hardwood forests of the Allegheny Plateau. Journal of Ecology 78: 443-458.

Whitney, G.G. and W.C. Davis. 1986. From primitive woods to cultivated woodlots. Thoreau and the forest history of Concord, Massachusetts. Journal of Forest History. April 1986: 70-81.

Whittaker, R.H. 1965. Dominance and diversity in land plant communities. Science 147: 250-260.

Whittaker. R.H. 1969. Evolution of diversity in plant communities. Brookhaven Symposia in Biology 22: 178-196.

Whittaker, R.H. 1975. Communities and ecosystems. Macmillan Publishing Company, Inc., New York. 385 pp.

*762-2
5-44
CC
S

DATE DUE

DUE DATE SUBJECT TO CHANGE IF A RECALL IS REQUESTED

DEMCO, INC. 38-2931